The COT Planning Guide

Second Edition

Tips, Tactics, and Strategies for Successful IC Outsourcing

Anthony Simon

and

Ronald Vogel

Library of Congress Control Number: 2002090226

ISBN: 1-931541-98-1

Printed by Lightning Source, Inc., La Vergne, TN.

Published by Simon Publications, Safety Harbor, FL.

Dedicated to Arbella Issabey, Keegen Vogel,
and Reiko, Stephanie and Christina Simon, and our parents
whose support enabled us to complete this book.

Acknowledgments

The authors would like to thank the efforts of those whose contributions have made this book possible. Special thanks to Marjan Kashvad, Kelly Lo, Mark Bapst, Teresa Stewart, and the folks at Simon Publications for their contributions and advice.

Table of Contents

Prelude

The first edition of The COT Planning Guide came out in the fall of 2001. The response and interest in the book took the authors by surprise. Many companies approached the authors through their website (www.GoFabless.com) and asked if they could participate in upcoming editions of The COT Planning Guide.

This second edition includes more information on design services; assembly, package, and test vendors; flip-chip designs; and other critical areas of the integrated circuit (IC) outsourcing industry.

In addition, several chapters include a section called "The COT Perspective" where we provide articles directly from experts in various fields of semiconductor design and manufacturing. These articles are packed with advice and information for those working on COT projects.

Our goal is to make this the definitive source of advice for managing and planning a customer-owned tooling (COT) semiconductor project.

Of course, in an industry as dynamic as semiconductor design and manufacturing, a single definitive volume would be impossible without frequent updates and improvements. So because the semiconductor industry changes so rapidly, and because of the popular demand for this guide, the authors have decided to publish this guide biannually.

Market Review

In the first three quarters of 2000, demand for IC outsourcing overwhelmed the industry's ability to meet capacity forecasts. Capital expenditures were on the rise and everyone was proclaiming the sky was the limit. Early 2000 estimates expected semiconductor revenue to hit greater than $300 billion. Almost any start-up semiconductor company with an idea was getting millions of dollars in venture funding. During those fat times, the only goal was to build a team, get funded, build a bigger team, get the architectural design started, and wait to get acquired.

What a difference a year makes.

Final year 2000 semiconductor revenues came in around 220 billion dollars with fabless semiconductor companies generating close to 17 billion dollars. In 2001, total semiconducotr sales dropped about 31 percent to 151 billion dollars while fabless semiconductor sales dropped 24 percent to 13 billion dollars.

Some economists believe we have hit the bottom, but it is still too early to determine how long the bottom will last. Economics will dictate whether many of the early start-up companies will live or die. Although capacity is abundant, IC vendors are scrutinizing companies more diligently and are now requiring full disclosure of business plans to help validate the viability of the start-ups. Both existing and new start-ups are finding it more challenging to obtain venture capital funding.

Of course, this situation is not new in our roller-coaster industry. Figure 1 illustrates the cyclical nature of semiconductor industry revenue growth over the last 25 years, during which there have been three major dips, in 1975, 1985, and 2001.

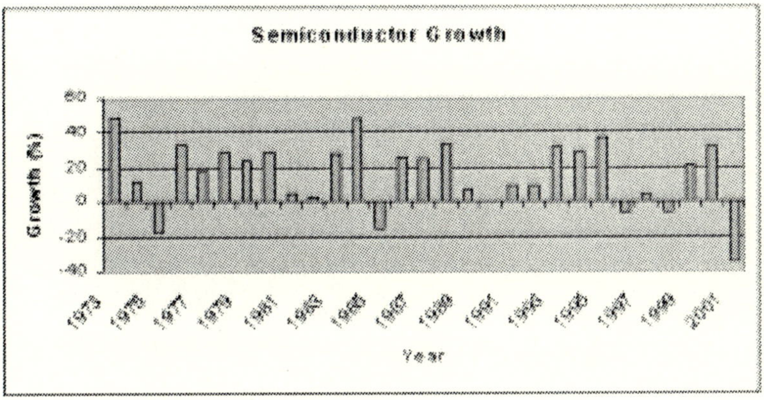

Figure 1: Semiconductor Industry Revenue

There are signs that the economy as a whole is picking back up. If history repeats itself, we should expect semiconductor year-on-year revenue improvement in 2002.

Developing a comprehensive IC outsourcing model together with a solid business plan will help you succeed in both good times and bad. In addition, whether we are in an upside or a downside situation, the relationship you establish with your vendors and your ability to achieve your objectives through careful evaluation of your suppliers will make the difference between success and failure for your business. We hope that this guide will help you achieve good IC outsourcing management and a successful business.

CHAPTER 1

The COT Model

This book is designed to serve as your guide if you are contemplating a fabless semiconductor start-up, or you are working with a company that has decided to develop its own integrated circuit (IC).

By "going fabless," you have decided to subcontract IC manufacturing to someone else. There are several ways to do this, but for most applications, the most cost-effective and technically efficient fabless model is the customer owned tooling or COT model. Unlike the application-specific integrated circuit (ASIC) model, where you rely on the layout and manufacturing capabilities of an ASIC vendor, the COT model is about controlling your own destiny, working directly with pure-play foundries, and setting up your own supply chain.

COT is exciting because it is finally leveling the playing field between fabless companies and the older integrated device manufacturers (IDMs) in a way that previous fabless models never could.[1] However, COT is a new concept to many. Also, there are some subtle differences between the COT model and older fabless models that many in the industry may not be

Figure 2: 300 mm Wafer (Courtesy of UMC)

accustomed to. This unfamiliarity with the COT model has led to many bad decisions, with results ranging from simply frustrating to occasionally catastrophic. The purpose of this book is to help you avoid making bad decisions and to help make your COT experience a glowing success.

The differences between the COT model and other fabless models are important. From the foundry perspective, customer-owned tooling means that the customer supplies the complete database of a chip design from logic to layout, generally in a GDSII format. This subtle difference has tremendous implications.

1 An integrated device manufacturer is a product company with internal foundry resources. ST Microelectronics, Texas Instruments, and Intel are examples of IDMs.

By supplying the complete database of your design to the foundry, you are, in essence, taking on the responsibility for manufacturing your chip. This is the glory of COT. You not only take complete control of how your chip is designed, but also have the ability to directly control the costs and business success of your project.

However, there are many issues to be aware of. For example, in order to succeed, you must know where and how to find intellectual property (IP); how to ensure the IP is validated for a given foundry; how to develop a successful relationship with a foundry; how to develop a test strategy to ensure cost-effective production; how to set up and manage back-end services such as packaging, assembly, and testing; how to handle the logistics of shipping chips; and how to manage supply and forecasting for your product lines.

If you have ASIC experience or have worked in larger IDMs, you may have seen many of these services handled by internal organizations. The details of COT manufacturing therefore might be new to you. It is easy to underestimate the extent of your responsibility for each step in the manufacturing flow.

This book has been put together to light a path for those entering the COT world for the first time. Understanding the implications of the COT model and avoiding the pitfalls is key to success in a COT project.

Why Go Fabless?

Even in down markets, fabless companies still grow faster than IDMs. The Fabless Semiconductor Association (FSA)

Figure 3. Wafer Acceptance Test (Courtesy of UMC)

estimates that by 2010, 50 percent of all wafer manufacturing will be outsourced directly to foundries.

These statistics indicate that "going COT" is more than just a fad. It represents a major shift in the semiconductor industry. One reason is that the pure-play foundry model, by focusing on wafer manufacturing, is proving to be successful. Foundries are even outpacing IDMs, long known for providing leading-edge technologies. In fact, the pure-play foundries have consistently surpassed the Semiconductor Industry Association road map for process introduction since the time of 0.25um complementary metal oxide silicon (CMOS) technology process. As of this writing, foundries are in production with 0.13um technology.

One result of the pure-play foundries' success is that IDMs themselves are starting to use foundries strategically as key suppliers of the most leading-edge technologies and capacity.

The availability of foundries has made an even playing field for IDMs and fabless companies to compete on. Fabless companies now have access to cutting-edge process technology, and they can focus more on product development and marketing. In order to compete, IDMs need to devote significantly more resources to process technology development. The price tag for a new 300mm foundry is approaching $4 billion. Working directly with pure-play foundries in the COT model allows you to be on the cutting edge of process technology without having to purchase it yourself.

The COT model also has advantages over other fabless models such as ASIC and FPGA vendor models.

FPGAs have the advantage of reduced time to market: the implementation of a design in FPGA can go very quickly since the FPGA is programmable. The disadvantages of FPGAs in many applications are high cost per chip, high power consumption, and insufficient performance. For many high-volume/lower-margin consumer applications, FPGAs have historically not been able to meet cost, power, and performance requirements.[2]

The ASIC model is convenient but very expensive. Typically, ASIC vendors take responsibility for a chip's development at the gate level, performing back-end design services and manufacturing the chip. Their fees include non-recurring engineering (NRE) expenses as well as a hefty unit price per chip. With this

2 At the time of this writing, FPGAs are still too expensive and power-hungry for high-volume/low-margin applications. That said, it is clear that the FPGA model has potential for these markets.

extra service, the per-chip cost can be higher than manufacturing the chip with a foundry.[3]

In addition, from a pure design standpoint, COT offers some tremendous advantages for today's highly integrated ICs. Today's chips are integrating the kitchen sink, including analog functions, embedded memory, and processors, giving rise to the concept of "system on a chip" or SOC.

The knitting of multiple subsystems together in a SOC is complex. Knowledge of how the subsystems will work in the complete design is required in order to complete the total system architecture. You cannot separate the front-end logic design from the back-end design. The COT model allows you to harmonize your front- and back-end design efforts. The ASIC model, on the other hand, by definition dictates a division between layout and design engineering, which can impede the time it takes to finalize a circuit. This added time could kill market opportunity.

The SOC designs of today, because they integrate so much variable functionality, require most designers to seek more third-party intellectual property (IP) than ever before. SOCs will help drive the growth opportunity in the IP market. The COT model gives you more flexibility than other models to use the best IP available for a given function. In the ASIC model, you are limited to the IP validated by the ASIC vendor.

3 Recently, the fabless ASIC model has been introduced to the industry. At the time of this writing, it is too early to say if this model will be successful.

Another nice feature about COT is IP protection. Pure-play foundries are focused on the business of manufacturing wafers, not selling products. You do not have to worry about a foundry becoming your competitor. This is in sharp contrast to IDMs. Even ASIC vendor companies often sell their own standard or semi-standard products.

It is also worth noting that the COT model may be the only option available to some companies, especially start-ups, since ASIC companies have begun focusing their efforts on supporting only top-tiered companies. Still, you are more likely to get support from a foundry if you have a solid business plan.

At the end of the day, the COT model is superior because it is the most cost-effective, technically flexible model for much of the IC business out there. In order to win in today's chip business, you need to get a product out quickly to meet decreasing time-to-market opportunities, protect margins by minimizing design and manufacturing costs, and stay ahead of the competition through design differentiation while protecting your IP. All these challenges are best met through the COT model, where you control your design from start to finish, you control your IP, you control your design costs, and most important, you control your design cycle time.

Meeting COT Challenges

The industry shift to COT has given rise to many new challenges. This book will help you meet these challenges and enlighten you about COT processes.

The COT model is all about planning and taking responsibility. In the COT world, you are in control. For example, you must manage foundry relations, IP selection, yield management, quality, and reliability. This book particularly addresseş those issues where we consistently see new entrants into the COT arena suffering from the most problems and misconceptions.

To address these issues we have organized the book in the following way:

Chapter 2: **The Project Plan.** The development of a project plan is critical given the complexities of a COT project. A good plan will serve as the outline of success for your IC program as well as providing a budget for your project.

Chapter 3: **Intellectual Property and Design Services**. IP is addressed early because it is where we see the most problems. This chapter will highlight IP selection, licensing, and manufacturing issues. It will also discuss the advantages of design services and offer tips on vendor selection.

Chapter 4: **Wafer Foundry**. This chapter will guide you in selecting a foundry and making your foundry relationship a success.

Chapter 5: **Packaging, Assembly, and Testing**. No IC manufacturing is complete without a package, assembly, and test process to complete the ICs. This section highlights how to select the right back-end vendors.

Chapter 6: **Design Tapeout**. This chapter explains the basic steps needed to tape out a device with a foundry.

Chapter 7: **Qualification and Characterization**. While the optimization of process technology is still a function of the foundry, in the COT model it is your responsibility to make sure your chip is yielding at the highest possible rate. This chapter focuses on how to efficiently manufacture reliable products.

Chapter 8: **Planning and Logistics.** This chapter helps you get through your first couple of chip projects without buying expensive software, by showing you how to manage the flow of chip production to meet your forecast and backlog.

Chapter 9: **Cost and Pricing Analysis**. This chapter will illustrate how to calculate chip production costs.

Chapter 10: **Flip-Chip Considerations.** This chapter discusses issues related to flip-chip technology, which is certain to rise in importance to the industry. Flip-chip designs represent a very small part of the total volume of commercial CMOS chips manufactured today. However, with ever-increasing pin counts and speed requirements, flip-chip technology promises to become a mainstream technology. This chapter outlines some of the manufacturing and planning logistics associated with flip-chip products.

You may be tempted to focus on selected chapters, but we advise you to read the entire book prior to starting your project, unless you have experience in COT foundry manufacturing. Although we are forced to physically structure the book in sequential pages and chapters, that sequence does not form a step-by-step recipe for your project. You will find suggestions in each chapter that may affect your planning from the beginning of your project.

CHAPTER 2

The Project Plan

An important first step is to put together a project plan. COT manufacturing involves the orchestration and cooperation of many individuals and companies. A good plan will help you manage and budget for these resources. Your project plan should include a summary of your objective and an estimate of the resources and budget needed to reach that objective. In addition, the project plan should address any high-level technical issues that are critical to success.

Using this top-down approach will help you identify your IP, foundry, assembly, package, test, failure analysis, and logistics process specifics. In addition, you will get a flavor for the resource allocations and tools necessary to support the design.

Start with the End in Mind

You start your plan by crystallizing your objective. The objective is the "what" and "when" section of your plan. From a high level, you list the major elements of your IC design and estimate when you need this design for prototyping and

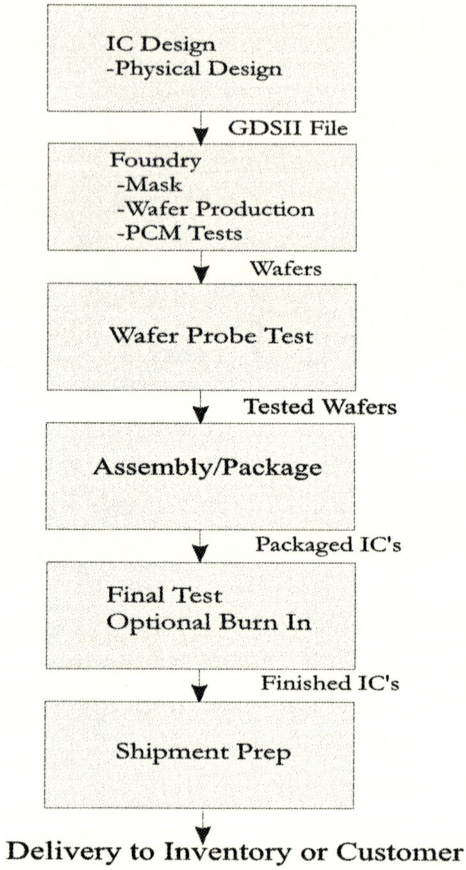

Figure 4: Typical Manufacturing Process

production. From a design standpoint, the elements of this objective should include the process technology you are planning to use, the estimated pin count, significant IP blocks, and the estimated gate count, die size, and power consumption. It should also list clock speeds, package requirements, and memory requirements.

Create a time line by planning the dates for prototype delivery and production requirements. Include time for silicon revisions and re-spins. A sample objective summary, time line summary, and design summary are shown below.

Sample Objective Summary

> The project objective is to manufacture the XR4576 network device in the 0.13um process and deliver prototype samples by 1/1/03 with production delivery by 10/15/03.

Sample Time Line Summary

Estimated Tapeout Date Revision A	11/15/02
In-House Prototype Revision A Samples	1/01/03
Estimated Tapeout Date Revision B	3/15/03
In-House Prototype Revision B Samples	5/05/03
Characterization and Qualification Complete	8/01/03
Production Parts Delivery Start	10/15/03

Sample Design Summary

Process Technology	0.13um, 1P8M, 1.2V/3.3V
Pin Count	1100 I/O pins
Package Type	FC-BGA
Gate Count	5M
SRAM Requirements	4mb
Power Consumption	< 10 W
Intellectual Property	
Serdes	3.125Gbps
LVDS	1.25Gbps
LDT	800MHz
PCI-X Interface	133MHz
Standard Cell Libraries with JTAG Support	
PLL (4X multiple)	1GHz with 0.1 UI jitter
Memory Requirements	
256Kx 8 DP SRAM (6 instances)	
2 Mb SRAM w/Redundancy (1 instance)	
128 x 64 Rx/Tx FIFO (4 instances)	

Table 1: Sample Design Summary

Resources and Budget

Once you have outlined your core objective, you can start to think about how to reach those goals. The "how and where" part in a COT project includes the resources, technical issues, and budget planning details that, together with your stated objective, become your project plan.

The "why" part of the equation is beyond the scope of this text. Why you are executing an IC manufacturing plan is the subject of your business plan.

To complete your project plan, start by highlighting the resource and budget elements required for your IC manufacturing process.

Personnel

IC projects involve the participation of a large group of engineers and specialists. A typical design effort will include the following internal and external resources: system architects, chip designers, software engineers, system integration engineers, layout engineers, test engineers, and engineering managers. In addition to engineering, you need to account for marketing resources, as well as legal and contract specialists.

For budget planning, it is good to use a spreadsheet to calculate total labor costs. By entering the man/months expenses per quarter, a spreadsheet approach eases tracking of total personnel expenses for the project. The same spreadsheet can also be used to track other investment costs, as discussed below.

Intellectual Property and Design Services

Given the complex nature of SOC designs today, your project will require the use of third-party IP. From a resource standpoint, finding, evaluating, and negotiating third-party IP can be a time-consuming process. You will need a competent team who can undertake these functions. IP negotiations always take longer than expected. Based on the IP you think you will need, make the best estimates to fit your project plan.

In some cases, you may also need to hire design resources to develop certain custom blocks for your design or to manage your whole back-end design. These resources also need to be considered in your project plan.

A good estimation of IP budget impacts is a necessary part of your project plan. IP license fees are potentially huge expenses. And negotiating IP, especially high-level IP, can become quite complex and time-consuming.

Once you have your IP, depending on your circumstances you may need to outsource some or all components of your design flow. You may want to consider this option if your company is a start-up, or if your company does not have much chip design capability or resources. Selecting a competent design service partner can be critical.

EDA Tools and License Costs

EDA tool costs can be significant, especially if the most recent high-end tools are required. Matching EDA requirements with your needs and budgeting for tool license purchases are therefore important planning steps. In addition, consider the hardware costs of running these tools. Can you get away with low-cost Linux machines or do you need the latest and greatest? While hardware usually is a smaller expense than software licenses, as you add engineers software licenses can add up.

Foundry Expenses

The major foundry expense to plan for is the mask development charge. This charge is very much a function of process technology; older technologies are cheaper than newer technologies. Mask pricing can range anywhere from $40K for a

0.35um mask set to over $500K for a 0.13um mask set. These costs become significant, especially if you plan to have many silicon revisions using a leading-edge technology.

Remember to plan for both prototyping stages and production. For a prototype, you will need to plan for at least one metal revision unless you are supremely confident. Planning for revisions will help avoid budget overruns.

Packaging, Assembly, and Testing

Once your wafers are completed by the foundry, you need to set up a supply chain to sort, package, and test them. From a budget and planning standpoint, each stage in the IC build-up can impose up-front NRE fees and potential lead times.

From a test standpoint, you need to plan for expenses such as device-under-test (DUT) and load-board development. The test program development itself is also an expense that should be accounted for. Once the device is available, account for expenses needed to characterize and validate the chip for production.

Also, depending on the package you intend to use, there are implications that should be accounted for. For example, BGA substrate design costs for higher-end SOC chips can reach $50,000.

Flip-chips will require additional costs due to added manufacturing steps. Flip-chips also have some special test and failure analysis challenges that are described at the end of this book.

Production

Going into production requires characterization and qualification of your chip. Account for the test resources necessary for each step. These resources are also likely to be in charge of making sure that your production yield is optimized for a given process technology.

Failure analysis capability is another requirement. No semiconductor IC is without its bugs. Having the proper failure analysis resources ready to support you in both prototype and production phases is critical to your success.

Logistics

Good logistics guarantees smooth-running process flows, from shipping to forecasting to inventory control and RMA procedures. Service vendors such as a shipping company need to be set up in advance of your project start and should be part of your resource planning.

Technical Issues

In addition to resources and budget, your project plan should outline high-level technical issues that could stand in the way of success. COT projects can become quite complex. Meticulous planning to deal with technical issues will help you avoid unnecessary time delays and expenses.

Typical issues you may encounter include finding the right EDA tools to match your design objectives; the right IP to work in your chosen foundry; memory that meets your performance requirements; libraries and IP that support your EDA views; and fuse repair and testing associated with memory redundancy. (New techniques may eliminate the need for fuse-repair external testing.)

Your plan is complete once you have tied your stated objective to your resource and budget plan and addressed the high-level technical issues for your COT project. After you complete this plan, you will be able to compare its results with your company's business plan to make certain the project is in line with your financial and strategic goals.

Major Points of Chapter 2

- Start your COT endeavor with a well-thought-out project plan.
- Begin your planning by outlining your design objective.
- Plan the resources and budget that will be required to reach your objective.
- Highlight major technical issues in your project planning.
- Use a spreadsheet to help track your total investment in head count, development costs, and other investment costs and purchases.
- Once you have finalized the project plan, compare it with your company's business plan to make certain the financials comply with your business strategy.
- There are many opportunities to sub-contract work in a COT project; consider these possibilities in your COT project strategy.

CHAPTER 3

Intellectual Property and Design Services

There is a surprisingly high amount of confusion about the use of third-party intellectual property (IP). We consistently run into experienced engineering managers who are eager to start their designs but are totally unaware of the challenges of finding critical IP blocks for their designs.

Some of the reasons for this confusion have to do with the dynamics of the COT model. COT demands that you take total responsibility for the development, layout, manufacture, and test of your design. COT is unlike the ASIC model, where you basically complete your HDL then throw the rest of the project over the wall and wait for chips to come back. In the COT model, you are responsible for "the whole enchilada."

In this chapter, we try to dispel some misconceptions we consistently encounter about IP, as well as offering advice on how to seek out and select IP blocks for your designs. Tomes could be written about IP use; our objective here is to focus only on those aspects that could affect your design flow and bottom line.

Major Concerns

Despite what you might have heard, in many ways the IP market is still very much in its infancy. There is no doubt that this market is developing and expanding quickly, with standards evolving to help ensure the ease of IP re-use. However, keep in mind that a good deal of the IP that is out there may not be validated or even available, despite marketing hype. Therefore, when looking for IP blocks for your design, be suspicious and never assume anything.

Hard and Soft IP

Generally speaking, IP is lumped into two categories: hard IP and soft IP. Hard IP refers to products such as digital-to-analog converters (DACs), analog-to-digital converters (ADCs), and phase lock loops (PLLs). Soft IP refers to synthesizable logic independent of process; it tends to serve unique applications such as embedded processors or advanced compilers. Soft IP is generally supported by readily available tools and has had a proven track record in silicon. Multiple third-party software drivers for soft IP cores may also be available on the market.

"Foundry IP"

One consistent misconception with developers and project managers is the notion that IP available from a foundry is actually designed and tested by the foundry. This is not the case. Ninety-nine percent of what is labeled as "foundry IP" is in fact third-party IP.

This may be confusing; after all, foundries often advertise availability of baseline IP libraries. At first glance one would simply assume that these libraries were developed by the foundry themselves and are as safe and solid as a savings bond. This is not exactly the case. While foundries are working very diligently with IP vendors to improve the availability and reliability of IP products, seamless "validation-free" design reuse is not here yet. (This issue is a hot industry topic. Everyone will stand to gain from an efficient design reuse model.)

Foundries are able to provide libraries free of charge because they actually pay the royalties themselves to third-party IP providers. Foundries introduced free-of-charge libraries during the time of the 0.35um process. Since that time, foundries have been expanding the breadth and sophistication of their library offerings.

Custom Solutions

Typical foundry libraries are developed to support a wide variety of customers, but if you are interested in a design with very low power or very high performance, you will probably be better off using a custom solution. Many vendors offer custom solutions that include standard cells, I/Os, and memories.

Your first inclination might be to go with a mainstream public company to get your IP, but that might limit your options. Many private vendors offer exceptional silicon-proven solutions. These vendors tend not to spend significant resources on marketing their products themselves but instead get by on "word of mouth." Yet these private vendors will give you a reference list that contains the "who's who" of the semiconductor industry. For example, Dolphin Technology is a private company that offers high-performance solutions in all the major foundries' leading-edge technologies; it was the original development partner for the Hyperlink Transport I/O for Advanced Micro Devices.

Single or Multiple Vendors

When you are choosing a vendor to support your high-performance requirements, it is nice if the vendor offers a wide range of silicon-verified IP solutions, so you can integrate the various IPs with confidence. Also, you can potentially reduce the overall price tag by bundling the IP package. For example, your design may require GPIO with variable drive strengths and slew rates, a custom CAM, dual-port and single port memories, HSTL, SSTL, PCI-X, and a Hypertransport core. If you have to go with multiple vendors, you may get stuck with large non-recurring engineering (NRE) expenses and royalty payments, as well as IP that might be incompatible and hard to integrate. Look for a vendor that will also give you the commitment of support, pricing, and time lines to support your design needs.

Validating IP Blocks

As an incentive for IP vendors to validate their IP blocks on foundries' processes, foundries offer to run test silicon through their fabs via multi-project wafers (MPW). (For more discussion on MPWs please refer to Chapter 4.) This practice is beneficial to both the foundry and the IP vendor. By validating IP blocks with a foundry, the IP house can guarantee that its designs are proven in silicon. Foundries, on the other hand, want to validate with as many IP vendors as possible in order to universalize the foundries' process technologies.

The fact that foundry IP is third-party IP has two major implications. First, you need to validate that the IP cores offered by the foundr y are not only technically compliant with your requirements, but also up to date with your foundry process technology evolutions. Remember, process technologies are not standardized across foundries. Therefore, IP needs to keep up with your specific foundry evolutions and design rule changes.

Second, unless the foundry provides a particular piece of IP for free, you will have to negotiate license terms and conditions for the IP directly with the IP vendor. In short, the responsibility falls on you to validate and in some cases negotiate terms for IP with third-party IP vendors. But here is the most important point: the foundry is not responsible in any case for third-party IP. (Some vendor's simulations could vary significantly from the silicon models!) Foundries also work very closely with EDA vendors to provide accurate parasitic and RC delay models, but the IP has to be proven in silicon to be certain.

In the turnkey ASIC model, designers tend to worry less about these issues. You simply tell the ASIC company what you need. For example, a reasonable request to an ASIC account manager might be: "I need a 120MHz microprocessor, three downstream 16-256 QAM blocks with FEC, and embedded 10-bit 40 MHz ADCs." The account manager would nod and present you with an NRE and per-chip quote, and you would receive your chip 9 to 12 months later. The ASIC company takes the responsibility to validate its process with the IP blocks, and it guarantees functionality and compliance with its process technology. So if there is a problem, they take responsibility to fix it.

In contrast, if you walked into your foundry's account manager's office and told him you needed the blocks mentioned above, he would take a different approach. The account manager would direct you to IP vendor partners. The foundry would also be able to provide a list of hard IP libraries that were developed to service a large number of users in terms of density and performance. However, it would be up to you to determine if the level of validation performed by these IP vendors was sufficient to guarantee compatibility with a particular process technology.

IP Technical Issues

When considering IP from a third party, you need to ensure that the IP is technically feasible. There are two major questions that need to be explored:

Is the IP block compliant with the process technology?

Is the IP block usable in your design methodology from both a hardware and a software perspective?

Compliance with Process Technology

First, you need to ensure that the IP block in question has been validated in some way, preferably in silicon. In addition, if the IP application requires third-party certification, proof of certification compliance may also be required. The risk associated with the IP is much lower when the block has been successfully integrated several times with other customers and has cycled through some volume production.

In assuring compliance, the IP vendor should also demonstrate that its IP blocks are keeping up with a foundry's design rule and SPICE model evolutions. This point is consistently forgotten, which can lead to migraine-sized repercussions when you are ready to go into production. For example, suppose you are designing for 0.18um process technology, where the dynamic voltage swing is only 1.8V, but the IP vendor designed and validated its blocks for 0.25um process technology, where the allowable voltage swing is much higher. If the vendor did not make the necessary modifications for 0.18um, you could have some severe difficulties.

Of course, these issues do not affect soft IP cores, where you synthesize to a given process technology. The world is moving toward soft IP, which has some obvious advantages, including faster time to market, the ability to port to multiple processes, and the fact that the IP is guaranteed by simulation.

A problem associated with soft IP is that the IP provider may not have simulated or validated all situations where the IP can be used. Therefore, you can encounter a bug never before seen. The more a block of soft IP is used, the smaller the probability of this happening.

Usability in Design Methodology

To address usability, you have to look at the deliverables provided by the IP vendor. For example, in what form is the block provided? Is it VHDL, Verilog, or GDSII? If the vendor is providing VHDL blocks, you need to make certain these blocks are not only compliant with industry-leading tools, but more importantly compliant with your own design methodology.

A block's usability also has to do with integration into a design. One must ensure that the IP block contains synthesis scripts that include compilation and template constraint files. In addition, it would be nice to be able to test the core after it has been integrated into a design. Testability is a weak point in today's world. If the IP core is very complex, the IP vendor had better be prepared to offer high-coverage test vectors or ATPG and scan scripts together with the block.

From a software perspective, it is important to know what will be required to exercise the block in silicon. In developing your drivers, it would be useful to have the source code associated with any microcode or sample driver code that your IP vendor has developed. In addition, you should ensure that no special requirements exist for simulating the functionality of the block in a dual software and hardware mode.

This raises the issue of embedded processor cores. It is a good idea to make certain any selected processor is well supported by tools and operating system software. Most of the well-known processor IP companies are well supported by multiple third-party sources for tools such as compilers, and work on various operating systems and applications software.

As mentioned above, foundries are now offering basic memory compilers as part of their "free of cost" library offerings. Various compilers are available, such as single- and dual-port register files. In addition to these foundry-exclusive compilers, third-party vendors also offer memory compilers that can be licensed, such as memories with redundancy.

Typically, foundries provide silicon-proven bit cells to the IP provider who develops the compilers. Some IP providers develop their own and therefore need to get it signed off if it does not meet a foundry's design rules. Bit cells usually violate rules to optimize performance and die size. These bit cells need to be well characterized by the foundries to meet specifications and yield. If a third-party vendor uses its own bit cell, ensure that you get recommendations and reports if possible on production products. Always make certain the foundry supports the bit cell in question.

Vendor Comparisons

From a technical standpoint, your evaluation of IP will not always focus on one vendor. Most likely you will face situations where you have to choose between competing vendors for similar IP blocks.

A good approach to decision-making is to tabulate the feature sets of competing vendor IP blocks for the types of blocks you are evaluating. Creating this decision matrix allows you to compare competing vendor IP features all at once, which will aid you considerably in finalizing your decision.

To illustration these decision matrices, we provide some templates for popular blocks below. These feature lists are for illustration purposes only. Your own list may be more detailed for your specific design environment.

IP Vendor Comparison Tables

Feature	Vendor A	Vendor B	Vendor C
Library costs			
Optimization priority			
Metal layers used in library			
Maximum density (Kgates/mm 2)			
Cell height (routing tracks - um)			
Vertical routing pitch (um)			
Horizontal routing pitch (um)			
Number of functions/cells			
Number of drives/buffer and inv.			
Number of drives/cells			
Pass gate inputs			
Buffered clocks			
Buffered outputs			
Sync flops			
Flops have scan equivalent			
0 ns setup/hold			
Availability			
CAD DC and PT. support			
Verilog or VHDL support			
Verilog or VHDL models encrypted?			
CAD place and route support			
CAD LVS/DRC support			

Table 2: Core Libraries Comparison (Courtesy of Mark Bapst)

Feature	Vendor A	Vendor B	Vendor C
Library costs			
Cell height in line (um)			
Cell height staggered (um)			
Min pitch in line (um)			
Min pitch staggered (um)			
Drive voltage			
5V tolerant			
ESD (kV)			
Latch up (mA)			
USB 1.1 transceiver			
32kHz oscillator			
Bidi w/pull up/down			
Analog I/O			
Will do custom cells?			
Number of functions/cell			
Availability			
CAD DC and PT. support			
Verilog or VHDL support			
Verilog or VHDL models encrypted?			
CAD place and route support			
CAD LVS/DRC support			

Table 3: I/O Libraries Comparison (Courtesy of Mark Bapst)

Feature (*Table 4*)	Vendor A	Vendor B	Vendor C
Synchronous 1 Port:			
Bit cell size (um^2)			

Feature (*Table 4*)	**Vendor A**	**Vendor B**	**Vendor C**
4kX16 size (um^2)			
4kX16 standby power			
4kX16 active power (/MHz)			
Memory BIST support			
Byte write?			
Design rule clean			
Metal layers used			
Compiler cost			
Availability			
Synchronous 2 Port			
1kX16 size (um^2)			
1kX16 standby power			
1kX16 active power (/MHz)			
Memory BIST support			
Byte write?			
Design rule clean			
Metal layers used			
Compiler cost			
Availability			
ROM Compiler			
8kX16 size (um^2)			
8kX16 standby power			
8kX16 active power (/MHz)			
Memory BIST support			
Byte write?			
Design rule clean			

Feature (*Table 4*)	Vendor A	Vendor B	Vendor C
Metal layers used			
Compiler cost			
Availability			
Diffusion or via			

Table 4: Embedded Memories Comparison (Courtesy of Mark Bapst)

Feature	Vendor A	Vendor B	Vendor C
Foundry			
Power consumption active (mW)			
Power consumption standby (mW)			
Input frequency range			
Output frequency range			
Multiplier options			
Jitter			
Loop acquisition time			
Duty cycle variation			
Voltage			
Cell size (Ksq-um)			
Availability			

Table 5: PLLs Comparison (Courtesy of Mark Bapst)

Once you have made a decision on the IP vendor you would like to work with, the next step is to validate that the IP meets your technical requirements. Never accept an IP based on the hand-waving promises of the IP vendor salesman.

Be prepared to ask questions that will reveal the suitability and availability of the IP for your design. A technical question strategy is provided below to help you design your own list of questions.

Third-Party IP Questionnaire[1]

ASIC Design Details

Silicon Verification, Solution Completeness, and Physical Information

- Has the IP been implemented in silicon? Did it work?
- Has the IP been verified within a system environment (with software, etc.)?
- What was the performance by process?
- Is the design reusable?
- Has the IP been previously embedded in other designs?
- Are special I/Os required?
- How much software is required?
- Is software provided? What format?
- Are there any requirements for system completeness?
- What is the IP physical size? Power requirements?
- Does the design contain any special cells?

Embedded Memory Requirements

- What is the methodology for manufacturing test (built-in self-test (BIST), multiplex (mux) isolation, etc.)?
- Are test circuitry and/or vectors included?
- Functions (words, bits, ports, timing, enables, etc.)?

Synchronous Design

- Is the design synthesizable?
- Is the design static-timing verifiable?

1 Courtesy of Mark Bapst.

- Is the design scan testable?
- Is there only one active clock edge?

Design Robustness

- Does the IP contain any of the following:
    - Latches?
    - Pulse generators?
    - One-shots?
    - Tristates?
    - Cross couples gates?
    - Derived resets?
    - Delay lines?
    - Layout-dependent asynchronous timing?
    - Internally generated/conditional resets?
    - Layout-dependent ASIC timing?
    - Combinational feedback?
    - Other problem logic?

Block Initialization and Reset

- Are asynchronous or synchronous resets used (or both)?

Clocking

- How many clock domains are required?
- What is the frequency of each clock?
- Are there duty cycle restrictions?
- Are there any external chip clock dependencies?
- How many clock domain crossings are handled?
- Are there internally generated clocks?
- Does the design use only a single clock edge?
- Are there gated clocks?
- Are there requirements for skew or insertion delay matching to other clocks?

Internal Interfaces

- Is there a bus or co-processor interface to a processor (e.g. APB or VSIA)?
- Is the interface a slave or master?
- Is an external DMA required?
- Does the interface require bidirectional pins?
- Is special interrupt handling required?
- Are all pins and outputs registered?
- Are there other interface issues?

External Interfaces

- Any special (non-CMOS) electrical requirements?
- Is the functional definition clearly documented?
- Is the timing clearly documented?

Testability

- Is the design scannable?
- Is the scan already included in the design databases?
- If no scan, are manufacturing tests included?
- Are embedded memory tests included?
- What is the expected fault coverage of the block?
- Are there built-in debug structures?
- Is the design IDDQ-friendly?

Test Bench

- Is the test bench reusable at the system level?
- What procedures exist for testing in the target application?
- Was code coverage used? If so what was the coverage?
- Is the test bench self-checking?
- Are system models (SPW, MATLAB) required and available?
- Are all dependent models provided?

Soft-Core Deliverables

RTL

- Is the RTL synthesizable?
- What is the format (Verilog or VHDL)?
- Is the HDL encrypted? If so, explain.
- Is the HDL source RTL or gates?
- Is the RTL documented?

Test Bench

- What is the format (Verilog or VHDL)?
- Is the test bench self-checking?
- Is the test bench commented?

Specifications

- Is there a test document that describes the test bench?
- Is there an implementation specification?
- Is there an integration manual?
- Is there a user's/programmer's manual?

Database Installation

- Instructions or script?
- Describe synthesis scripts.
- Describe scan and automatic test program generation (ATPG) scripts.
- Describe static timing script tools.

Hard-Core Deliverables

Verilog or VHDL Sim Model

- What is the format?
- Sign off accuracy? Annotatable?

Test Bench

- What is the format (Verilog or VHDL)?

- Is the test bench self-checking?
- Is the test bench commented?
- Is there a test document which describes the test bench?

Physical Database

- Which EDA tools are supported?

Specifications

- Is there an implementation specification?
- Is there an integration manual?
- Is there a user's/programmer's manual?

Database Installation

- Instructions or script?

Synthesis Model

- Which EDA tools are supported?
- Is the timing model accurate?

Scan Support

- What is the test compiler/test generator model?
- Are the ATPG vectors provided, and if so, in what format?
- Does the HDL include a scan wrapper?

Static Timing Model

- What tools are supported?

EDA Tool Information

Tools Version

- What simulator and version were used for the design?
- What synthesizer and version were used for the design?
- What scan insertion tool and version were used for the design?
- What ATPG tool and version were used for the design?

- What static timing tool and version were used for the design?
- What formal verification tool and version were used for the design?

System Design Details

System Specifications

- To what system specification and version has the IP been designed and to what degree does the block meet the specification?

Performance Specification

- Against what performance specification and version has the IP been verified?

System Model

- Does a floating or fixed-point "C," etc., model exist for the IP?

Software Design Details

Language Portability

- What language is the software written in? If assembly language, what processor is it targeted to?

Software Requirements

- What is the MIPS requirement/estimate?
- What is the code space requirement/estimate?
- What assumptions are used for resource estimates (cache, memory bandwidth, memory size, etc.)?
- What dependencies are there in the OS?

IP Business Issues

IP business issues normally revolve around two topics: support and pricing. Do not assume that the greatest challenge in obtaining third-party IP is ensuring that the technical issues are met. It is highly likely that most, if not all, of the technical solutions you will need for your design are available today somewhere in the IP market. In many cases, therefore, the complexities of contract negotiations can be much more time-consuming than the technical due diligence. The greatest challenge in obtaining third-party IP is negotiating the use and price of this IP to ensure success for your business.

Contract Negotiations

Make certain you allow enough time for IP contract negotiations, which always seem to take up more time than you expect. In fact, it is a good idea to have your lawyer start reviewing the vendor's terms and conditions as soon as you start looking at an IP vendor seriously.

Remember also that in today's environment, there is no such thing as one-stop IP shopping. Almost every piece of IP out there will need some modifications or tweaks to fit your design requirements. Do not go into an IP negotiation with the notion of changing your design to meet IP requirements. Instead, define exactly what kind of IP and support you need to make your project successful. Write these requirements in a specification and use this as a starting point in IP vendor negotiations. Require the IP vendor to explain to you how they will fulfill your requirements. Providing a requirements specification from the beginning will also speed up the overall negotiation, because you will know exactly what to negotiate for.

Support

The support and price you negotiate with the vendor are key. Remember–it's almost never about the technology. It's about the support and price. Lack of support is more than a potential source of frustration. Quality of support will define who wins and who loses in the race to get a product to market. So spend a lot of time understanding how a piece of IP will be supported. For example, is there an applications engineer available for answering your questions on problems you face during integration? Will your IP vendor have a vested interest in responding in a timely fashion? What is the turn-around time for answers to critical questions?

In your contract negotiations you should clarify how support of the IP will be handled in the event the IP vendor is merged with or acquired by another company. This question goes beyond simply signing an agreement of support. For example, what happens if in the middle of your IP transaction, your direct competitor buys out your IP vendor? Even if you had a commitment of support, how can that truly be guaranteed or policed if the new company wants to put you out of business?

There are ways to protect yourself. One way is to agree that the IP source code is provided to you in the event of nonperformance, so that you can have it to port to new process technologies going forward. You might even list your competitors in the contract and agree that if one of them buys out your IP vendor, all contract-related IP is turned over to you. This clause may also help deter a potential competitor buyout. Whether an IP vendor will agree to this depends on how badly they want your business. (If you are a smaller company, you may not have the leverage to negotiate these terms, and in any case may prefer to focus on price and time to market. The last thing you want to do is lose a market window because you spent an inordinate amount of time negotiating.)

Of course, things may work out the opposite way. For example, your company may be bought out or taken over. Make sure you have assigned rights to the IP license under this scenario. In other words, make certain that liability is not one-sided.

References

Remember to check out your IP vendor. Get references. Call the references and make certain they are happy with the vendor's performance. If the vendor purports to have a relationship with a foundry, contact that foundry and confirm the relationship.

Measurable Milestones

You should also ensure that all IP transactions and contracts are based on measurable milestones which conform to an agreed-upon time line. These milestone triggers should include the final stages of your design—working silicon—and payments should be triggered according to these milestones.

Warranty Language

The contract should also incorporate warranty language and state clearly what should be done in case of failures of the IP. Many times, in complex designs, the root causes of design failures are not immediately clear. At what point will your vendor take responsibility for failure analysis? If root-cause failure is found in the IP, what do the terms and conditions say about getting the problem fixed? What if the failures show up once you have shipped your product to your customer? The terms and conditions should be clear on these points.

Be aware that there is no one model for buying IP. Like many things in life, IP license terms and conditions come in various shapes and sizes. Some standard cells such as basic I/O are provided free of charge. In these cases, the IP design files and GDSII files are available along with the source code for free.

Price Negotiations

Price negotiations require care and planning. Constantly consider different scenarios. In your contract you should specify which products or services are free and which are not free. For example, you may assume that software drivers for the IP are free of charge, only to find out that they require separate licensing or royalties. Are there maintenance fees associated with the driver software? What if there are new revisions of the software—are these free or chargeable?

Royalties and Up-Front Fees

As a general rule of thumb, it is always good to avoid royalties. Royalties come out of cost of goods sold and affect your gross profit margin. Royalty payments also add complexity to your accounting. Licensing fees, in contrast, come out of operating expenses and only affect bottom-line margins. So unless you have some extreme business case, avoid royalties if possible. If you cannot avoid them, try to minimize their impact by negotiating for lower royalties, or by trying to cap royalties in time. You might offer to invest some of your resources for co-developed IP that can come to you without royalties or even license fees.

Often you can negotiate higher up-front licensing fees in exchange for a lower royalty. In cases like these, present and future value calculations can come in handy to determine the total cost of the IP over time.

Up-front fees have other advantages, such as amortization. Furthermore, you can often negotiate the terms of payment for up-front fees. This is particularly handy when there is some work to be done by the IP developer. You can easily tie up-front fee payments to deliverable milestones. This provides incentives to the vendor to get his work done. It also minimizes your risk and allows you to pay out the up-front fee in smaller chunks over the lifetime of the contract.

Up-front fees are also easier to handle from an accounting aspect in terms of calculation and management. Royalty arrangements usually call for some type of audit mechanism. To be frank, however, with high-level IP it may be difficult if not impossible to avoid royalties.

Royalties can be calculated in various ways: as a percentage of the sales price of the end product, as a percentage of the profit margins of the end product, or as a percentage based on the IP size as a function of the total die size per product. (Die-size-based royalty has the disadvantage of rewarding the IP vendor for sloppy work, since the bigger the IP the more the royalty.) Royalties can also change over a period of time or as a function of volume. Work with your financial manager to come up with the best royalty scheme, in order to minimize financial impact to the per-unit cost of your chip. The number of payment models and terms is limited only by imagination.

As a general rule, always try to link any payment, whether it is a royalty or a license fee, back to either the performance of the vendor or the protection of your contract. Pay close attention to all obligations written for you and the licensor.

Single-Use License

Clarify whether your contract is a "single use" license. If you negotiate a single-use license, you may have to pay the license fee per design iteration. The definition of what constitutes an iteration is also a negotiable term in a contract. For example, is porting the design to a smaller process technology considered a new design? What if you have to make small changes or feature additions to a design and re-tapeout a chip? Does that constitute a design iteration? These terms need to be clarified and agreed to in a contract.

Deliverables

There are also questions about deliverables. We have seen many instances of start-ups negotiating for IP but forgetting to define exactly what is to be delivered in the contract. For example, will the IP be provided as a GDSII file or is it RTL- or HDL-level only? (If it is in HDL, it will be your responsibility to lay out the block in your chip.) You should also define acceptance criteria for the delivered IP and clarify time limits on how long you can use the IP. What happens if there is a limit? What are the terms to extend the license?

Other Strategies

Be proactive in suggesting solutions if your negotiations reach an impasse. For example, if you are asking for a lower price, always provide a creative mechanism to offer something in return, such as a solution to a problem. Often simply saying that you need a lower price is not an effective approach.

If you have multiple IP negotiations running in parallel, try to get the same terms for each block of IP. This will greatly simplify the management of your IP contracts once you are in production.

Realize that IP negotiations are not just about obtaining technology—they are about your company's commercial success. Ask yourself what you need to make your company successful. Make a list. Then make a list of all the contingencies that could harm your endeavor. Based on these lists, ascertain that your final contract provides the support and prices you need as well as the protection you need from contingencies.

Question Strategy

As you can see from the previous discussion, IP business negotiations can be complex. Take time and negotiate carefully. To help you avoid the most common pitfalls and get the best deal for your chip, we have put together a list of questions that you can think about or raise.

Vendor Support Questions

- How many other customers is the vendor currently supporting?
- How many people are in the company?
- What is the ratio of design engineers to applications engineers? For example, if you have ten people doing RTL design but only two people porting, will they really be able to do the design?
- Who will be available to support your design?
- Are there any large IDMs paying big money to your vendor that would limit the vendor's ability to support your design?
- What about future contracts or commitments that may be coming up? What if your design is not ready for some time? Will the IP vendor still be available to support you in the future?
- Will the vendor maintain support for changes in EDA methodology or other tool revisions?

- Once the IP is bought, what is the time line available to ensure technical support for the design?
- Where and in what time zone are the people who will provide the support?
- Have you negotiated turn-around or response times for both critical issues and less important issues?
- How is support provided? Phone call? Meetings? Email? All of the above?
- How is support handled in the case of failure in a design? How will failure analysis responsibility be handled between you and the vendor? (Keep in mind that in almost all cases of failure, the root cause is not known until failure analysis (FA) can be completed.)

IP Compliance

- How many processes has the vendor proven in silicon?
- Has the vendor proven with your chosen foundry and the target process technology?
- Are there any products available in volume that use that particular piece of IP?
- Is the IP vendor familiar with your foundry's latest technology process, design rules, and SPICE models?

License Negotiations

- What happens to your support if your IP vendor is bought out, merges, or goes bankrupt?
- What guarantees ensure that the IP vendor delivers as promised? Are payments based on clear milestones? Are there penalties the vendor must pay if certain milestones are not hit? In the event of nonperformance, do you have the ability to take and finish the design somewhere else?
- What is the cost model—is the IP single use, multi-use, or owned? Do you have to renegotiate each time?
- If you buy it, what exactly do you get?

- Have you negotiated for warranties, termination rights, and assignability?
- If there are changes in process technology or other design or manufacturing methodology, will support be affected?

Design Services

Foundry Engagement Models

Within the COT model, several foundry engagement models can be used by fabless companies. The most common is for the customer to interface directly with a foundry. Here a customer hands off a GDSII file to a foundry and directly manages the manufacturing process flow.

A second engagement model involves working with a design company to manage the foundry contact from design to physical prototyping.

A third and new engagement model involves the use of a fabless ASIC company. This is similar to a classic ASIC company except that the ASIC vendor takes responsibility for production and back-end support through a third-party foundry. (At the time of this writing, the fabless ASIC model has not yet been proven. It may provide a cheaper alternative than the classic approach.)

The focus of this book is the COT model. However, you should be aware of the options. Each engagement model above has both advantages and disadvantages. The decision as to which model to use is a function of your resource constraints and business situation.

Outsourcing

There could be a host of reasons why you might opt to outsource part or all of your design. Your resources may be constrained, or you may be involved in technology in which you do not have a lot of experience. Or you may be going through a chip project for the first time and feel better about having a helping hand.

With the complexity of ICs continuing to track Moore's Law[2] and consumers continuing to demonstrate an insatiable demand for "more" out of their electronic devices, the ability for companies to provide all elements of the IC value chain has rapidly diminished.

Advantages of Outsourcing

The argument for IC outsourcing starts at the design phase. With every two generations of process technology, a substantial change in IC physical design tools and methodology is necessary. The approaching shift to 0.13um will prompt the next major change. The good news is that 0.13um process technology will offer nearly twice as many equivalent gates as 0.18um. The bad news is that it will drive design toward a fully hierarchical methodology; require the use of larger blocks and new design-for-test (DFT) methodologies to deal with the inclusion of SRAM, flash, EDRAM, and analog blocks on chips; create new signal-integrity and power problems; and substantially increase the size of the team required to design ICs. All of

2 Moore's Law: The logic density of silicon ICs (the amount of information storable on a given amount of silicon) roughly doubles every year or 18 months.

these factors will significantly increase the cost of building and supporting an IC design infrastructure.

In addition to increasing the cost of the design infrastructure, the newly found real estate will also mandate the move of functionality onto chips at an increased rate. This creates a secondary effect. A single company will not be able to amass the required expertise in a timely, cost-effective manner to produce all of the IP required to populate the available sockets in leading ICs. In addition, it is increasingly difficult for companies to justify spending money to design standards-based IP. Companies will increasingly focus their precious resources on developing technology that will provide them competitive differentiation.

After a quick glimpse at design infrastructure, IP development, fabrication, test, packaging, and characterization, it's easy to understand why there are compelling cost reasons to contemplate outsourcing to design services companies.

So with such compelling arguments for IC outsourcing, why isn't everyone outsourcing the majority of the process today?

Disadvantages of Outsourcing

One obvious reason is time. Widespread outsourcing in any given segment of the electronics supply chain usually catches on slowly. Initial customer-supplier relationships are typically very tactical in nature, driven by short-term necessity rather than strategic intent. In fact, vendors that emerge in segments of the supply chain that have not been traditionally outsourced usually face significant market resistance in the beginning. The slow development of the Electronic Manufacturing Services (EMS) market is a perfect example. It has evolved from being a cheap labor pool for low-value manufacturing to the preferred model for the industry.

Additional reasons for not adopting IC outsourcing have included: a) lack of credible alternatives for IC physical design services; b) finding IP; c) business instability of IP suppliers; d) process compatibility issues with IP blocks; e) inability to gain access to necessary wafer volumes; and e) complexity of establishing and managing supply chain relationships with design service firms, library suppliers, IP vendors, foundries, packaging vendors, assembly and test houses, etc. All of these have historically been valid issues. However, IC outsourcing is maturing to the point that the problems associated with these issues are manageable.

The big remaining obstacle is accountability. Vendors in each segment will stand behind their individual work. However, none will guarantee you working silicon that meets your design, cost, and volume requirements in the end. Each component of the IC product development value chain has emerged as an independent island. Therefore, the big risk is that when you encounter problems, none of your vendors will accept full responsibility to solve them. Finger pointing is inevitable. This issue existed at the infant stage of the EMS market as well. Fortunately, the sophistication of the IC supply chain is making it easier for you to take full ownership and control of your design project. Vendors are more willing to share in the debug process to get you to volume production.[3]

3 Contributed by Jim Douglas, VP Marketing, Tality Corporation.

THE COT PERSPECTIVE

Ten Keys to Successful Design Outsourcing

Jim Douglas
VP Worldwide Marketing
Tality Corporation

For companies of all sizes, outsourcing has become an increasingly strategic tool. Innovative companies can accelerate elements of their product-development process to leading design firms in order to gain a competitive advantage.

A well-chosen design service partnership can help your company focus on core competency tasks that enable market differentiation. Such a partnership can also accelerate development schedules through access to special expertise and by reducing ramp-up time on new projects.

We've laid out the top ten ways to establish a successful working relationship with an external design firm.

1. Clearly define your requirements.

Since time-to-market is critical for companies, clearly defined requirements are crucial. Decide whether you want a design firm to augment your current project team, take on a specific portion of the project, or take responsibility for the entire design. The more specific your requirements prior to engaging a design supplier, the better the end results will be. In today's high-speed, short life-cycle product environment, more and more companies are using external design firms as their engineering departments—with the external firm providing turnkey product and infrastructure development. This model has proven successful, and has led to the development and commercialization of many new technologies.

2. Develop a short list of potential firms.

Experience is an extremely important factor in selecting a design partner. The firms on your short list should have extensive experience in your product area and knowledge about associated technologies. Look for broad experience in a variety of technologies—with experience that goes deep into each area. Depth and breadth of experience indicates a firm's fundamental understanding of state-of-the-art technologies.

Evaluate the supply-chain partners each firm can offer. Can those partners potentially provide your company with value from concept through final production? Finally, look at the firm's key people. Do they understand your marketplace? Do they have experience successfully launching new products? Affirmative answers to these questions are imperative for every company on your short list.

3. Determine if you have a business model fit.

It's important to determine whether the firms on your short list have a business model that suits both your short- and long-term needs. The ideal partner will have experience operating under different models and be flexible enough to adopt a model that meets your specific requirements. Five common models you will encounter are listed below.

Scoped Time and Materials

The supplier provides an estimate of the man-hours required to complete the project and a rate per hour. After agreeing on scope and rate, the supplier works until the project is completed. Changes to the original statement of work will be documented and agreed on via an engineering change order and billed at the same hourly rate.

Fixed Price

The supplier defines the scope of work and fixed price to deliver the project. Changes to the original statement of work will be documented and agreed on via an engineering change order. Payment for engineering changes can be based on either a fixed priced or time and materials (as defined by the original contract or by subsequent negotiation).

Defined Level of Effort

The customer contracts a specific amount of manpower or man-hours over a specific period of time and identifies projects that resources will be used on. The supplier must ensure resources are available when required by the customer. Subsequently, the customer will be billed for unused time.

NRE Plus License Fee

The supplier provides engineering services, using one of the previously described business models. Additionally, the customer pays a one-time license fee for IP owned by the supplier incorporated into the final work product. The license fee provides the customer the right to use the IP in a specific field of use.

NRE Plus Royalty Stream

The supplier provides engineering services using one of the business models described above. Additionally, the customer pays a recurring fee (typically based on unit sales) for supplier-owned IP used in the final product. The fee could be based on best efforts or on some predetermined minimum volume target.

4. Evaluate project management competency.

Project management can make or break a collaborative design effort. Strong project management is a science, not an art. Ask prospective firms to articulate their project management processes and the standard practices you can expect for your project. Meet your prospective project manager. This person will be your key contact during the duration of the project. The supplier will measure them by successful completion of the project (to spec), customer satisfaction, and adherence to internal project financial targets. They are your best leverage to ensure that the project meets expectations.

5. Determine adaptability.

Product-development projects rarely, if ever, proceed without changes or unforeseen circumstances. Your requirements will likely be modified midstream based on incremental market intelligence or availability of required components. You may miss your partner's deadlines and they will, inevitably, not execute to perfection. Make sure you understand how your partner manages engineering changes and establish an escalation path that can be exercised if necessary.

6. Understand design infrastructure.

In addition to solid design expertise, one of the key advantages a potential design partnership should provide is a robust design infrastructure. The expense and time associated with building and maintaining a strong design infrastructure shouldn't be underestimated.

The firm should also interface efficiently with other organizations (internal and external) in your supply chain. Make sure you understand how they function in this area.

7. Determine quality and standards policies.

A defined quality and standards policy is essential. Look for a well-documented quality assurance program. Examine the company's design processes and its review system. Find out if and how the company tracks quality results—remember, continuous improvement requires continuous tracking.

8. Understand security policies and practices.

With so many sticky intellectual property (IP) issues swirling around the industry, it's important that a firm demonstrate whether it can keep your IP confidential. Does the firm take non-disclosure agreements seriously? Ask specifically about security policies and practices associated with client engagements. You must be certain the firm you choose is completely professional with regard to these issues.

9. Plan ahead and be ready to commit.

One of the key advantages of working with a design partner is the speed with which you can build a project team. Priority conflicts can sometimes lead to months-long delays when you attempt to form a project team with internal resources. However, don't expect the companies on your short list to be at their best if you haven't made decisions. They may not be able to start immediately, hold resources open indefinitely while you contemplate whether you are going to engage, or maintain a fixed delivery schedule, if you delay the start. You should plan your time carefully and be mentally committed to an efficient evaluation process. When you reach agreement on these issues with prospective firms, the process can move efficiently.

10. Get references—and check them.

Again, look for demonstrated expertise in the areas of interest to you. Ask for references from current or former customers. Then call each reference and ask for a detailed critique of the firm's performance. Ask specific questions about your particular concerns.

The strategic outsourcing of design services can help you meet your objectives and give you a competitive advantage. However, a well-planned approach—including careful analysis of contract resources, documented project planning, and continuous coordination throughout the life of a project—is essential.

Tality Corporation

Major Points of Chapter 3

- Foundries do not develop their own IP except in some limited circumstances. Independent IP vendors provide 99 percent of the libraries you find at a typical foundry.

- Ultimately you are responsible for ensuring that any third-party IP you use, whether or not it is provided by the foundry, is fully validated in the target process technology.

- You may have to obtain high-level IP on your own from IP vendors.

- In many cases you might have to negotiate payment terms for IP. Ensure that you negotiate with support in mind.

- When negotiating for IP, it is a good idea to be paranoid. (Andy Grove told us that only the paranoid survive.) Consider any possible threat to the long-term support of your IP block, such as a buy-out of the IP vendor.

- When negotiating payment terms, remember that royalties and up-front payments affect your bottom line in different ways.

- Whether you pay up-front fees, royalties, both, or some other payment agreement such as stock warrants or providing your children as collateral, remember to always tie vendor performance to future payments (or non-payments).

- Depending on your resources and expertise in certain areas, you may wish to consider outsourcing part or all of your COT project to a design-services company.

- As with any vendor, take the time and due diligence to thoroughly evaluate potential design service vendors.

CHAPTER 4

Wafer Foundry

This chapter discusses the process of foundry selection and manufacturing interfaces. All chip manufacturing starts with the foundry; therefore, the proper management of foundry relations is key to any product manufacturing success. This chapter will outline the main issues you need to know to ensure you choose a foundry wisely and to anticipate the basic requirements for engagement.

Foundry Selection

Pure-Play Foundry or IDM

Choosing a foundry can be a very complicated decision. Although many foundries claim leading edge and available capacity, it is your job to make certain you receive the support and commitment from the chosen foundry to support your design-in and production ramp. The top three pure-play foundries are UMC, TSMC, and Chartered Semiconductor. Newer foundries include ANAM, 1st Silicon, Silterra, and many others. But the pure-play foundry is not the only game in town.

Foundry capacity is also available from traditional IDMs. Understanding your needs and the pros and cons of working with the various foundries can make or break a product manufacturing strategy.

IDMs

Perhaps the riskiest option is dealing with an IDM that has only temporary capacity in its captive foundries. The semiconductor industry is famously cyclical; hence capacity that is available today may not be available tomorrow. In addition, since this is not necessarily the focal business of an IDM, there is a good chance that the marketing, sales, and support infrastructure to manage a foundry business may be seriously lacking.

By definition, IDMs do not focus on foundry business. It is unlikely, therefore, that the personnel selected to manage an

Figure 4: Typical FAB Bay (Courtesy of UMC)

IDM's foundry business are the company's "A" team. Even worse, they may staff foundry service with a factory operations team that cares little or nothing about foundry customer satisfaction.

Another difficulty with IDMs concerns protection of intellectual property. Unlike foundries, IDMs may not have experience managing and protecting third-party IP. In addition, if theIDM sees value in your IP, they may try to bargain for rights or at least for a piece of the business. Alternatively, some semiconductor companies will grant capacity services only to those whose IP presents an interesting business case.

On the technical side, an IDM may not be very good at supplying design rule and SPICE model documentation and support to their customers.

Nevertheless, it could be in your favor to use the facilities of an IDM if you want the company to invest in your business, if you are not able to find capacity elsewhere, or if the price or service offering from the IDM is very compelling. However, in most cases, you will be better off taking your business to a pure-play foundry.

Pure-Play Foundries

Today, pure-play foundries are taking the lead in new process technology introduction ahead of IDMs. In addition, pure-play foundries will typically have libraries available for these new processes since they introduce technologies and libraries concurrently.

The pure-play foundry is "pure-play" because they do nothing but manufacture silicon for customers; they are not interested in selling their own semiconductor products. Handing over your valuable IP is not a concern. This charter motivates pure-play foundries to develop long-term committed relationships with customers, as well as to provide customers with good service and accurate up-to-date information on design rules and SPICE models. Pure-play foundries are also interested in working with third-party IP vendors to validate IP designs in the foundry's manufacturing process.

Pure-play foundries concentrate their investment and growth strategy on continuously increasing the capacity and technological advancement of their manufacturing capability. Pure-plays often offer the widest possible process technology options with some of the most state-of-the-art and efficient facilities.

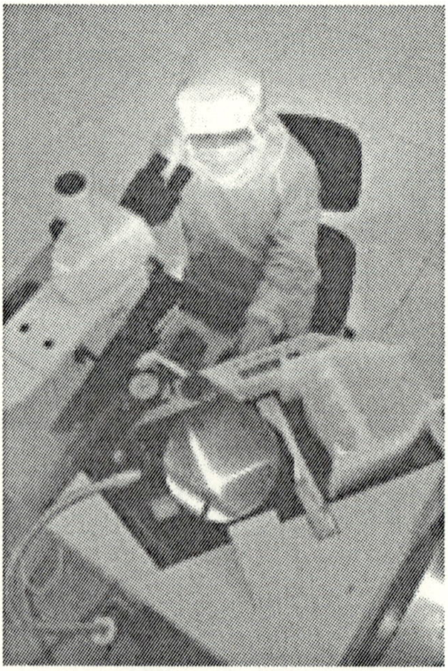

Figure 5: Wafer Inspection System (Courtesy of UMC)

In reviewing the options, it is most likely that you will want to work with a pure-play foundry unless business or technical circumstances dictate otherwise. In any event, the type of foundry is only one criterion. There are others, the most important of which are capacity, capability, price, support, and relationship.

Capacity

Determining a foundry's capacity or potential capacity is important if you intend to be in business over the long term. It is important to develop a trusting relationship and obtain commitments from a foundry, to help guarantee capacity when demand is high and everyone is scrambling for silicon capacity. However, if your foundry is not capable of meeting its commitments to you, no amount of relationship or good will can produce the chips you will need to meet your requirements.

Determining the present level of capacity of a foundry is not difficult, since the wafers produced per month by the foundry are fairly easy to determine. But you should also determine a foundry's capacity as a function of process technology. This information is typically available in financial reports. It is also important to look at the financials of a foundry, and determine if a proper amount of investment is being made in capital expenditures and in research and development.

Capacity is a function of manufacturing infrastructure technology as well as process technology. Have the candidate foundry show you their technology roadmap. Ensure it is keeping up with or exceeding industry-standard technology metrics such as effective transistor length, wafer diameter, and metal layers supported.

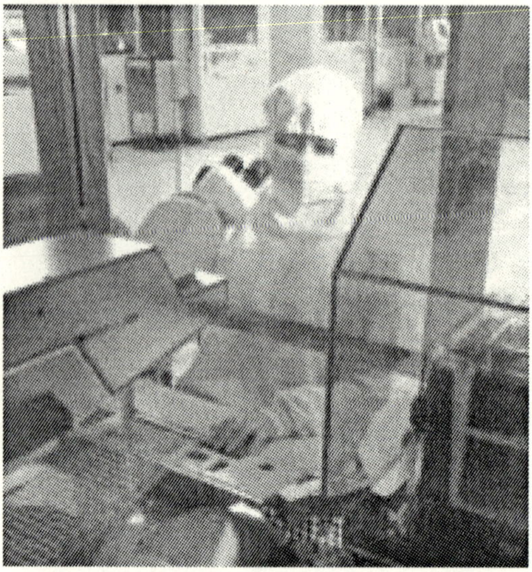

Figure 6: FAB Processing (Courtesy of UMC)

In addition, ask about the manufacturing technology and find out if the foundry's fabrication facilities are built using state-of-the-art techniques. For example, modular techniques help improve throughput by lowering defect densities per wafer. Also, state-of-the-art facilities use rail systems to move material through fabrication, improving cycle-time and yields. Another state-of-the-art technique is the standard mechanical interface (SMIF) system that provides a clean modular environment while transporting wafers: wafers are transported in wafer cassette holders and then transferred to pods that can be loaded onto equipment.

In discussing capacity, you should also ask what production management processes or technologies are implemented to help guarantee maximum efficiencies through the fabrication process. For example, is supply chain manufacturing tracked by computer, or available online for customers to track product going through the foundry? These options help immensely to make certain product is not side-railed or lost on its way through production. (Don't forget to ask foundries when they schedule their maintenance shutdowns. For Taiwan, this is usually around the Chinese New Year.)

Capability

Capability is a key factor that is related to the needs of your particular design. For example, if you are considering a flip-chip design, you want to make certain your foundry has developed close subcontractor vendor relationships to provide flip-chip capability.

More generally, a foundry should be able to brag about its breakthroughs in various aspects of semiconductor manufacturing. Beyond process technology and wafer diameter, there are also technologies concerned with new metal conductor materials, dielectric isolation between metal layers, and the number of metal layers supported through the process. You want your foundry to have aggressive road maps matched with a sizeable amount of success in meeting goals.

Relationship

As a start-up or smaller company, it may be your responsibility to convince a foundry that your prospects are rock-solid. Your goal is to develop a good relationship, so that the foundry will not only commit capacity, but also help you through the inevitable manufacturing problems that will arise as you go through the production process.

Unless times are really soft, do not assume that a foundry will automatically want your business. Like everyone else, pure-play foundries want to fill their fabrication lines with high-margin, high-volume, or preferably high volume *and* high margin business. Foundries will not want to waste their time on dubious or unpromising business opportunities.

Start-ups need to take particular care not only to establish but to strengthen good relationships with foundry partners. Go visit the foundry, meet with the business managers, present your business model. These investments in time and travel will pay off when the going gets tight and the economy is roaring.

Support

Support is important and ties in well with the notion of relationship. Both are necessary. Does the foundry provide online tracking of processes in the foundry such as yield, defect densities, financials, and wafer in process? Does the foundry provide electronic links and accept electronic tapeouts? (Note that pure-play foundries tend to be more open about these online tracking tools than typical IDMs.) Does the foundry use electronic links between sales offices and the production facilities in order to speed up tapeout and production start? Do not assume that everyone must support this, especially if you have decided to work with a smaller or newer manufacturing company.

Price

The price of silicon is a function of volume, design parameters, and process technology. It is good to have at least two or three foundries provide quotes for your business. Competitive quotes will make sure that you do not enter a relationship where prices are too out of line for your production. Although foundries quote on a per-wafer basis, keep in mind that a foundry that is better at optimizing a process technology and keeping defect densities low may in the long run be the cheaper option, even if their per-wafer price is slightly higher than the competition's. As defect densities improve, foundries can help you find the "sweet point," or the most cost-effective technology node, to design into for a given project. Similarly, foundries may provide various levels of building block libraries for free or preferred pricing, which may also affect aggregate price comparisons.

Location

Finally, while the location of the foundry itself is not too important, the location of its sales and support satellite offices could be a critical factor in its ability to support your design through production. While the major pure-play foundries are more than likely to have sales and support offices in locations around the world, the same may not be true for many ASIC companies or other IDMs.

A note about China: While mainland China's market is opening up to global trade, there are still questions as to China's policy on IP. At this time, there are no clear government policies regarding IP protection. In fact, some larger foundry customers do not allow GDSII data to be sent to China. These customers have their mask plates generated in another location. Always makes sure you are working with vendors with a proven track record of protecting your IP.

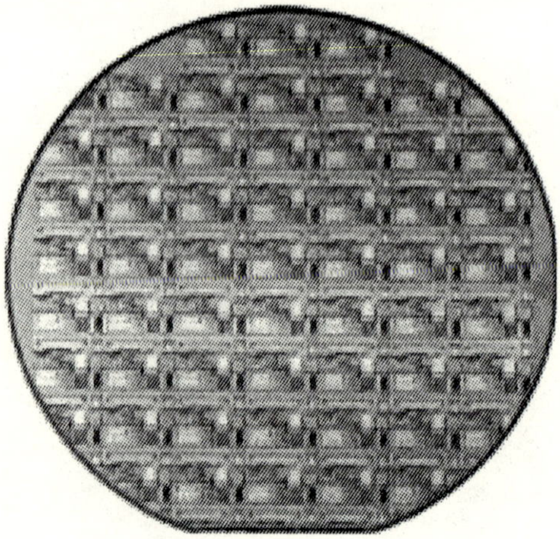

Figure 7. Multi-Project Wafer (Courtesy of IMEC-Europractice)

Other Points

It is an advantage to work with a foundry that owns geographically diversified fabrication facilities. You do not want an earthquake, political upheaval, or war to disrupt your wafer production. Also, keep in mind that well-known patents are held on semiconductor manufacturing. A professional foundry should be licensed under the portfolios of the major process technology patent holders including AT&T, Lucent, Harris Semiconductor, Hitachi Ltd., IBM, Motorola, Philips, Signetics, ST Microelectronics, the Lemelson Foundation, Standard Microsystems, and Texas Instruments.

A Special Note for Small-Volume Applications

Foundries like to run high volumes of a product because they are then better able to improve processes and reduce defect

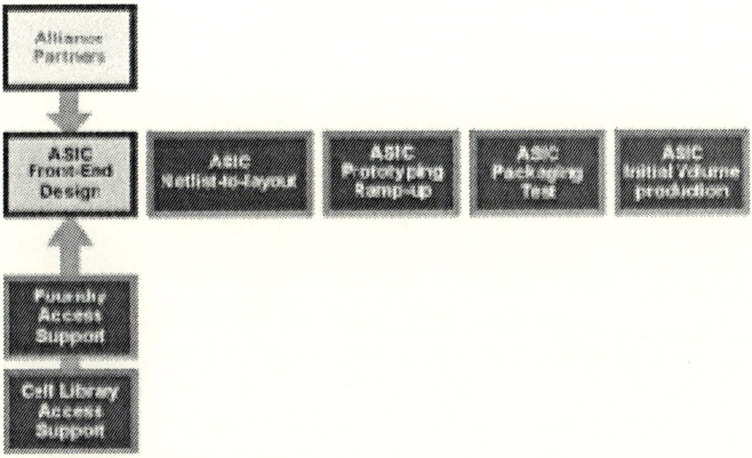

Figure 8: Prototype Process Flow (Courtesy of IMEC-Europractice)

densities by getting quick yield feedback from sort. If a lot scraps in fabrication, another lot can be started immediately with minimum impact on the customer. On the other-hand, if the foundry runs small quantities of a product, it is more challenging to improve yield and might require more resources. Also, a scrapped lot can have a major impact on the customer. Foundries will support start-ups and smaller companies only if they feel there is a financial or strategic return on the investment.

But many applications will never be high runners. If your foundry is not willing to support your design, you can use a prototyping service. Companies like IMEC-Europractice run multi-project wafers (MPW) and can develop a dedicated mask set when you need higher volumes and cost improvements. One nice feature about these companies is that you can get direct access to the foundries' libraries and technical data, and they will support full design-rule check (DRC), ERC, LVS check, and GDSII preparation. Essentially, they act as a one-stop shop. These prototyping service companies have partnered with leading foundries to offer a wide variety of mixed-mode and leading-edge technologies.

In addition, some prototyping companies can offer full turnkey services, including place/route, back-end sort, package/assembly, and test services.

Locking In

Assume you have done your due diligence and have decided on one or two foundries that meet your requirements.

If you are having trouble making the final decision between foundry options, you can develop a decision matrix with weighted categories in order to reach closure.

Getting a Quote

At this point you will want to get written quotes from the foundries that have made it onto your short list. The quotation process usually starts with an interview with a foundry's account manager.

A typical quote example is provided at the end of this chapter. In this example, the prices have been left out in order to avoid creating wrong expectations for pricing. Foundry pricing can change drastically as a function of technology process, industry demand, and other factors.

Description of Target Process Technology

A quote should start with a description of the target process technology. In our example, the quote is for a 0.35um process with five-layer metal, one-layer poly, 3.3 volt core, and 5V-tolerant I/O. In addition, the design will use 20 masks that will be used to create 22 photo layers to complete the design. This implies that one or two of the 20 masks will be reused during the photolithography process. This is important to know, because in an actual build, cycle times are calculated based on photo layers per unit time, not masks per unit time.

A great place to get wafer pricing by technology is through the FSA (Fabless Semiconductor Association), which surveys fabless semiconductor companies and generates quarterly wafer pricing reports. The reports include IDM pricing for comparison.

In addition to providing a basic description of the technology, the example quote states that the process will not include special additions such as anti-electrostatic discharge (ESD) protection, epi material, and polymide layers. [1]

NRE Pricing for Mask Tooling

A quote should also provide NRE pricing for mask tooling. Normally, masks are charged differently according to their grades. In the quote example below, the mask tooling charges and classifications are provided in an attachment. Normally,

1 Epi material is a highly resistive wafer implant used to reduce sub-threshold current. Polymide layers are used for alpha particle contamination reduction. ESD implants are used for extra circuit protection against electrostatic discharge.

wafers are classified in different grades that are priced according to their respective levels of complexity to build. For example, a wafer mask that needs to generate very tight geometries will be more expensive than a mask that needs to process only rough geometries. More expensive layers include poly layers, which define the actual gate level details.

In our example, "Alpha" grade refers to these gate-level poly layers. (Foundries may name their grades differently.) Gate-level lithography requires extremely tight geometries. Mask tooling for this layer requires a great deal of manipulation and use of special procedures. In contrast, the layer describing a pad ring is an example of a less complex layer. Relative to the rest of the circuit, pads have very large geometries averaging 40um to 60um.

The need to create semiconductors with tighter and tighter geometries as process technologies shrink is one reason why mask-tooling charges have been increasing exponentially in recent years. The mask layers that can generate these tiny geometries take time and extensive data manipulation. Mask development time can vary from a few days to a month. Obviously as these complexities and development times increase, so do associated costs. Total mask tooling charges are also increasing due to the fact that most complex designs today require more mask layers than ever before. From the attachment in the quote example below, we can immediately see the impact of more masks to the bottom line.

Lot Categories

In addition to mask tooling charges, the quote you receive from the foundry may also specify the relative costs of different lot categories. These categories, which are generic in the industry, are defined as engineering lot, hot lot, and super-hot lot. The quote example below explains the relative cycle-times among these categories. It is general industry practice to put a proto-lot for a new design into hot lot status. Foundries know that there normally is a lot of work to be done to get a design into production, and not much of this work can get started until their customers have silicon in their hands. Super-hot lots actually require that each stage in the production line stop and wait for super-hot lot wafers to come through. Due to the disruption that super-hot lots cause, they are not always available, even if you are willing to pay premium prices. This is especially true in times of allocation. Prices are normally two to three times higher for a super-hot lot than for a hot proto-lot.

Quotes will also point out the different prices associated with prototype lots and production lots. Foundries will normally build 12 wafers for a proto run and guarantee that six will meet an agreed-upon wafer acceptance test (WAT); since it is a proto-run, the foundry wants to cover itself for any potential fallout. Foundries normally quote 25-wafer production runs.

Post-Wafer-Production Services

Foundries are usually interested only in wafer production, since this is where they see their highest value. But if you are lucky enough to have a foundry agree to wafer probing service or other post-wafer-production services, the price of these services should also appear on your quote. Foundries normally offer these services only to customers that can maintain a large stable production volume through the foundry, in order to get immediate yield feedback into the fabrication line to improve defect densities.

Date of Validity

Finally, a quote will always provide a date of validity. Since prices are always changing in the semiconductor industry, foundries normally provide quotes that are valid for only 30 days. The terms and conditions on import and export usually follow Incoterms guidelines, an internationally accepted set of rules that outline how liabilities are shared on cross-border shipping.

The Written Contract

Once you have made the decision to work with a particular foundry, agree on a written contract or set of terms and conditions that will outline your mutual expectations for your relationships. (Larger foundry customers may have a separate contract, while mid-sized to smaller companies will operate under existing foundry terms and conditions.) As in most contract negotiations, you should always consult an attorney before signing any documents. This is especially true for a contract with a foundry.

The objective of any foundry contract or terms and conditions is to establish in sufficient detail the terms under which you expect to do business. At a minimum, the contract should address production procedures, pricing, delivery and order terms, reliability and quality guarantees, warranty and warranty limitations, and return materials authorizations (RMA). It should also address other legal issues such as dispute resolution, indemnification, licensing, exclusions, confidentiality (with a non-disclosure agreement (NDA)), and contract termination.

All the topics in the contract or terms are important. However, the production procedures, pricing, and delivery and order terms are going to have direct impact on your day-to-day business. For example, foundries often require their customers to submit demand forecasts over a specified period (usually six months). These forecasts are considered written commitments to purchase products over that time frame. Foundries will often allow leeway in this and specify commitments on a decreasing percentage of forecasted quantities over time. These terms will also outline a customer's ability to cancel orders, when to submit purchase orders, and other conditions. Make certain you have a good understanding of these terms and their implications before signing.

From a technical standpoint, the production procedures should outline what happens as a result of a design change notification on your part or a technology or material change notification on the foundry's part. For example, there may be foundry material notifications that could have impact on production. The rules for dealing with engineering change notifications should be clearly spelled out in the contract or terms and conditions.

The contract or terms and conditions should also specify the size of order lots for prototype, corner splits, and production. Typically a production lot is 25 wafers and a proto-lot is 12 wafers with a delivery guarantee of six wafers. The terms will also outline your ability to stop shipments and hold wafers as they move through production.

You should read the contract or terms and conditions with great care and with legal counsel present. Ask the foundry questions if you do not understand certain terms in the contract. If there is a term or condition which stands out to you as unfair or untenable, negotiate to have it changed. Of course, your ability to negotiate terms will be proportional to the strategic significance your business represents to the foundry.

You have worked hard to locate and decide on a foundry partner. You chose a foundry based on its ability to support you in terms of capacity, capability, support, relationship, and price. Now you can concentrate on the next step, the tapeout process and production.

Foundry Forms

Quotation

1. Process Technology and Mask Description

A. 0.35um 1P5M 3.3/5V generic logic process.

B. 20 masks, 22 photo layers, non-epi starting materials, ESD implant and polymide layers not included.

2. NRE for Mask Tooling

Please see Mask tooling price table attached below

3. NRE for Prototypes

Super-Hot Lot (0.9 day/layer, limited availability): $$$

Hot Lot (1.2 days/layer): $$

Normal Lot (1.8 days/layer): $

Proto-lot size starts with 12 wafers, guarantees six to PCM.

4.Production Pricing

Wafer price per lot size of 25 wafers.

5.Payment Terms

Payment expected 30 days after shipment.

6.Validity 30 days.

The industry normally makes quotes valid for 30 days.

Form 1: Quotation Example

Mask Grade and Price Form

Level#	*Description*	*Mask Grade*
1	Diffusion	Beta
2	P-Well	Kappa
3	VTDV	Kappa
4	TG	Kappa
5	Poly	Alpha
6	VTNH	Kappa
7	N+	Kappa
8	P+	Kappa
9	SAB	Kappa
10	Contact	Beta
11	Metal1	Beta
12	Mvia1	Beta
13	Metal2	Beta
14	Mvia2	Beta
15	Metal3	Beta
16	Mvia3	Beta
17	Metal4	Beta
18	Mvia4	Beta
19	Metal5	Beta
20	Pad Window	Delta

Form 2: Mask Grade and Price Table

Mask Prices

Grade	*Mask Count*	*Unit Price(US$)*
Alpha	1	$$$
Beta	11	$$
Kappa	7	$$
Delta	1	$
Total	20	

Total Price= **1**($$$)+**11(**$$)+**7**($$)+**1**($)

Form 3: Mask Price Calculation

Major Points of Chapter 4

- Choose a foundry based on capacity, capability, price, support, and relationship.

- Make an extra effort to develop strong relationships with your selected foundry partner to ensure long-term support and allocation, especially during boom times.

- If possible, get competitive price quotations for your business from more than one foundry to ensure you are getting the right price.

- Understand the pros and cons of working with IDMs and pure-play foundries.

- Work with legal counsel when negotiating your contract with a foundry.

- Understand that your price quotation should include all NRE, prototype, production, and mask development charges for a given process technology.

- The price of a mask is a function of its geometric complexity.

CHAPTER 5

Packaging, Assembly, and Testing

Once you have selected a foundry to build your wafers, you need to consider how to get those wafers sorted, diced, packaged, and tested.

Most foundries are capable of providing these back-end (test and package) turnkey solutions, but they tend to shy away if the volumes are small. Still, the trend is for these services to be outsourced. In fact, by the end of 2002, more than 70 percent of all fabless semiconductor companies will be requesting test services from subcontractors specializing in testing. Therefore, it is good to understand how to select these vendors.

Vendor Selection

The typical manufacturing flow includes wafer fabrication, wafer probing, packaging/assembly, testing, and delivery. Planning is important to make certain that each subcontractor is working in harmony with the production flow and that necessary materials are ordered in advance or just in time for each stage of production. The number of subcontractors you choose to use in your process flow can depend on technical factors, costs, or logistical requirements. For example, you may be working with a high-pin-count package that requires special test equipment.

It is easier, from a logistical point of view, to work with one assembly house for all your back-end requirements. That way the foundry can drop-ship completed wafers directly to your assembly house, which then takes responsibility for wafer probe, package, and assembly services, wafer testing, and delivery to you or your customer. Most of the big-name subcontractors perform sort, assembly, and test services. Other companies, such as ANAM, ASE, ChipPAC, and SPIL, not only provide all these services but also have test houses set up in the United States to help with failure analysis and test development.

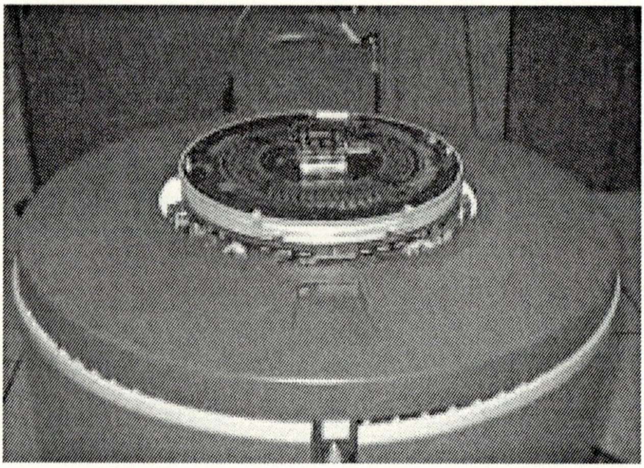

Figure 9: Loadboard (Courtesy of SABER)

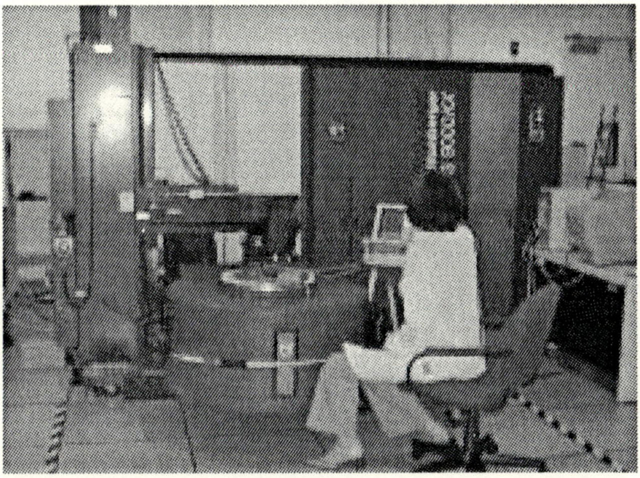

Figure 10: Schlumberger ITS 9000 ATE (Courtesy of SABER)

Whether you choose to work with your foundry, with a set of back-end vendors, or with a "one-stop shop" assembly and test house, choose your vendor(s) on these golden criteria: capability, geography, pricing, cycle time, capacity, and relationship.

Capability

Capability is a quality that takes a high priority. Ensure that the vendor(s) in charge of your test and packaging are capable of meeting the demands of your device. (In some cases, such as with flip-chip products, the actual assembly of the chip requires special capabilities as well.)

For testing, you need to make certain your test vendor has the right probe and automatic test equipment (ATE) for your wafer-level and final package–level tests. If you expect your design to have a very high I/O count or extremely high clock speeds, work with your prospective vendor(s) to make certain they can accommodate these requirements. Test vendors sometimes specialize in certain types of technology, like the testing of analog circuitry or memory chips. Obviously, if you are developing a high-pin-count, mixed-signal, multiple-non-integer clocked SOC, you want to seek out vendors who specialize in these types of testing.

It is a good idea to familiarize yourself with the major ATE manufacturers and learn their latest product offerings. Learn what ATE capabilities are out there for probe[1] and test services, and relate that information back to your vendor to help ascertain capability.

Test capability is not only a question of test equipment. It also requires experience in the development of test application boards such as loadboards or device under test (DUT) boards. In your evaluation process, ask prospective vendors if they have experience developing test boards for systems similar to your design. For example, if your design is a multifaceted SOC that requires certain at-speed tests, it is preferable to work with a vendor with experience in developing test boards for these requirements.

Capability also applies to packaging. Prospective vendors should be able to meet your package needs as well as provide any package-related services. Every design is unique, and yours

1 As designs increase in speed and input/output density, probe card technology is also improving to meet the needs of the higher performance and more complex designs. As I/O counts skyrocket, not only does the number of pads increase, but also the placement of pads is changing. Peripheral pads are migrating to the interior of the die. Probe technology needs to keep up by providing interior probe pins on a high-pin-count tester.

may require special considerations in terms of pin count, pitch, thermal and electrical requirements, and other factors. In addition, new packaging alternatives such as chip-scale packages should be available if you need them. Sit down with prospective vendors to make sure your specific requirements are supported.[2]

Geography

The geography of your back-end vendors is also a factor when setting up your manufacturing flow. Ideally, there should be excellent logistics between the back-end test, package and assembly, and final test vendors in order to optimize shipping efficiencies.

Geography is also an important factor in failure analysis. Some major test vendors have satellite offices in countries around the world that can provide failure analysis services. For example, test vendors headquartered in Taiwan and Korea will have offices and laboratories set up in Silicon Valley. Having local services can make a big difference when your design runs into trouble. Of course, you do not have to be limited to using your

2 A quick package primer: Generally speaking, two baseline package technologies are commercially used today, wire bond and flip-chip. Wire bond is the most common. As its name implies, wire bond technology connects die pads to electrical substrate by means of extremely thin, usually gold, wires. Ball-grid array (BGA), QFP, and DIP packages all use wire bond technology. Flip-chip technology, on the other hand, implements the use of solder balls to serve as the electrical connection from the active face of a die directly to the substrate. In today's world, flip-chip is definitely on the cutting edge of commercial package applications.

package/test vendor for failure analysis. There may be excellent vendors that perform these services in your area.

Pricing

The prices that vendors charge for services are also a key criterion in making your final vendor selections. It is always important to get multiple quotes from competing vendors to make certain your pricing is in line. Also remember that the prices you pay for per-chip test, package, and assembly have to be added to the per-chip cost of foundry manufacturing to obtain the total cost per chip. And the total cost per chip has a direct impact on your gross margins. Therefore, it is a good idea to seek the lowest prices possible for every stage of the manufacturing process without giving up quality and responsiveness.

Cycle Times

Probe prospective vendors about cycle times. Cycle times will vary depending on the package and test requirements for your design. As you survey competing vendors, you should develop a sense of what cycle times are acceptable. Imagine if you paid a premium to your foundry for a super-hot lot only to find out that your assembly house could not complete your circuit for months. Find out the rules for expediting products through back-end vendors. Make certain that your products will be eligible for expediting if and when you need to get products through quickly.

Capacity

You should ensure that your vendors can handle the volumes and demands for services that you expect. Capacity applies to you whether or not you expect to manufacture in high volumes. You want to make certain that a larger test house does not neglect your low-volume business in order to cater to someone else's high-volume business. Similarly, if you are forecasting extremely high-volume business, you probably want to stick with the larger vendors. Challenge your vendors by asking them about their throughput capacity and whether or not they can handle your forecasted business.

Relationship

Relationship is also a key factor when deciding on your vendors. Have you worked with a prospective vendor before? Was the experience a happy one? If you are interviewing vendors for the first time, it is important to judge whom you want to work with.

Once you decide on a vendor, you need to take the time to develop a good relationship with the vendor's managers and key employees. Just like foundries, back-end vendors can run out of capacity during periods of high demand. Like everyone else, they want to prioritize their time with the clients to whom they have forged commitments. These commitments are based on business and sometimes personal relationships.

Finally, in the vendor selection process, it is a good idea to get references on prospective vendors. Ask your contacts in the industry about their experiences working with various back-end vendors.

Time- and Money-Saving Tips

Develop a Test Plan that Emphasizes Test Efficiency

The goal of package and test planning is to minimize costs in production while achieving quality goals. A look at what is happening in the industry illustrates why this is important. As SOC circuits gain complexity, test equipment must be able to keep up. This race results in test equipment getting expensive, upwards of $5-$10 million a machine. Test vendors amortizing these high costs can charge you more than $300 per hour to use these expensive equipment.

About ten years ago, testing represented less than 10 percent of the total cost of manufacturing commercial ICs. Today this percentage can approach 40 percent for some high-end solutions. Load boards and DUT boards may incur additional NRE if your design has unique testing requirements.

There are strategies you can implement in order to minimize test costs without compromising quality.

Avoid Expensive ATE Test Equipment

One strategy is to avoid requiring the use of expensive ATE test equipment. For example, say your expected pin requirements will exceed 1000 pins. At the time of this writing, only the most expensive ATEs are capable of testing SOCs with 1000 pins. You can avoid these expensive ATEs by multiplexing test chains and test patterns through 500 pins instead of 1000. That way your device will be able to be tested on older, cheaper, but just as capable ATEs.

Minimize Test Time

If you cannot mux your I/O, and you need to use the latest and greatest ATE available, you can at least take the necessary design steps to minimize test time per chip. One way is to incorporate design for test (DFT), such as built-in self-test (BIST), into your logic design. Almost all SOC designs will require a certain amount of logic-phase test planning to accommodate the complexities of your design. The extent to which you incorporate DFT and BIST also has to be weighed against the impact that would have on the schedule and its effect on die size. The most popular DFT approach is to use scan chains throughout the design in order to speed the ability to test large parts of the chip in short order. You can use BIST for embedded memory as well as processors.

There are other common methods to limit test time. For example, you should strive to run multiple tests in parallel per die or chip. And if possible, run your tests at higher clock frequencies. Also, in order to minimize the total test time for a batch of devices, consider having multi-site load boards set up to run wafers or chips in parallel.

The test time for complex SOCs can be about 3 to 5 seconds per chip at wafer test, and 4 to 6 seconds per chip at final test. Remember that soak time for hot- and cold-temperature testing, as well as loading large vector sets, also adds to total test time.

Reduce Analog Test Costs

Finally, you can consider ways to reduce analog test costs. Often, due to the difficulty of mixed-signal circuits, companies like to run analog test chips prior to the tapeout of a SOC. One idea is to place analog test chips on a multi-project wafer, which reduces costs because the cost of a multi-project wafer is shared. See chapter 6 for more information on multi-project wafers.

Avoid Scrapping Packaged Product

Increased package costs are making probe testing much more important. Using an expensive package is like wrapping dollars around every chip—money you throw away if you need to scrap the chip. Therefore, you should strive to get as much coverage as possible at sort in order to sift out bad dies. Analog testing is still a challenge at the wafer level, but good fault coverage and DC testing should help make up for this.

On the other hand, wafer probe testing may not be required at all when the cost of packaging is a fraction of the overall production chain and the die size is small. Situations like this may justify performing all tests at final test. Once again, it is important to consider these options at the beginning of the design phase in order to ensure an efficient test strategy.

Beware of Package Costs

Packages that use a PCB laminate often will require design NRE to be paid in advance to a package design vendor. NRE may be a factor if any custom rework of the laminate is required for your design. The NRE for a custom BGA substrate design can cost anywhere from $10K to $100K, depending on complexity. If you have latitude in your design, try to keep your requirements as generic as possible.

This also applies to package choice. A myriad of packages is available from inexpensive DIPs to very expensive flip-chips. Package material also affects package price, with plastic being the most cost-effective and widely used material today. Ceramic packages fetch a good premium over plastic.

Expensive packaging can account for a great percentage of the per-chip manufacturing cost in many SOC designs. So it is important not only to design for test, but to design for package as well. On this note, you should work closely with package and assembly vendors early on (during place and route) to achieve optimal bond-out for your IC.

Plan for Package Development and Approval

Package development and approval has the following phases: design, customer approval, package development, and production cycle time. The duration of each of these steps varies widely depending on the package technology and strategy. For example, package design usually takes from ten days to two weeks. Customer approval can take two hours to one month, and development time to handoff samples can take six to eight weeks. Package standard cycle time in production is one to eight weeks. Qualification can be accomplished by using the vendor's existing data on similar packages.

Keep an Eye on Logistical Details

Administratively, the package assembly house will require bonding diagrams from you in advance. The vendor will need time to review and approve these diagrams. The vendor will make certain that your design does not violate assembly house design rules and cause bonding violations such as angles, crossovers, or bond pad pitch violations. Assembly houses keep a list of design rules for each of their packages. Needless to say, avoiding design rule violations can save you time and headaches in the long run.

Your sort test vendor will also need a wafer built up with pads in order to build a probe card or simply your design layout. As discussed in the next chapter, foundries usually run an extra wafer for this purpose at no charge. Ensure that the foundry knows to whom it should drop ship this wafer in advance.

Depending on the package and your relationship with the package vendor, you may need to order materials to supply to your assembly house. This point is often forgotten by many companies rolling through the IC manufacturing process for the first time. For example, for packages requiring a lead frame such as QFP or DIP, you may need to order the lead frame and arrange shipment to the assembly house well in advance.

Keep in mind that in periods of allocation, parts like lead frames may have lead times running into several weeks. PCB substrate is another example of material that can be hard to come by in periods of allocation.

In dealing with your back-end vendors, you will most likely need to specify certain services and tests that may seem trivial and obvious. For example, you will probably want the assembly or test house to perform lead scanning on your devices to check for coplanarity of the leads. Often this service will not be completed unless you request it.

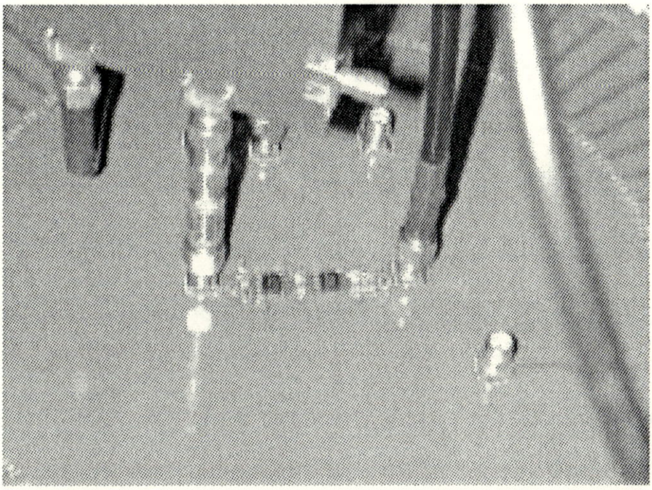

Figure 11: OC-192 Transceiver Text Fixture (Courtesy of SABER.)

Finally, ensure that each vendor knows to whom to drop ship product at each stage of manufacture. If you are using one assembly house for all your back-end services, your foundry should have drop-ship instructions for your assembly house. Similarly, your assembly house will need instructions on where to drop ship completed material. The logistics get more complicated when more vendors are involved. Supply chain management and logistics are further addressed in chapters seven and eight.

A Note about Subcontractors

Working with test subcontractors can help you reduce your overhead while fulfilling time-to-market requirements. Areas that you may consider outsourcing include test design consulting, debugging, and characterization. The functions you can outsource in each area are listed below.

Test Design Consulting

- Device under test (DUT) interface
- Automatic test program generation (ATPG) definition
- Test strategy
- DFT

Debugging and Characterization

Test Program Development

- Special acceptance (analog testing, critical path testing)
- DUT interface design and build
- Device characterization

Design Debug and Validation

- ATPG integration
- Silicon debug, diagnostic, and repair
- Characterization to production
- Test solution optimization
- Test solution deployment
- Probe handler interface

Once you have selected your back-end vendor, you will be asked to fill out a test questionnaire that will specify the complete testing and handling of your product. A sample test questionnaire is provided below. It outlines test issues such as vector size, memory tests, wafer sort, and burn-in. In addition, the test questionnaire asks for instructions for top-side marking, lead scanning, and packing and shipping.

Customer Test Flow Questionnaire

Customer: ______________________________

I. PROJECT TARGET DATES:

PRODUCTION TARGET DATE:____________

Estimated Wafer Tapeout Date:______________

Wafer Sort Setup Date:__________________

Assembly Completion Date: ________________

Final Test Setup Date : ___________________

IMPORTANT: Please submit transfer plan and production ramp-up plan with (work week/year or month/year) for capacity planning.

II. PRODUCT INFORMATION:

Product Name: _________________________

A. Package/Lead Count: ____________________

NOTE : Please submit list of products if number of products identified for transfer is more than ONE.

B. Product Application: _____________________

C. Functionality: *Logic* __ *Analog* __ *Mixed Signal* __ *Embedded Memory* __ *Linear* __ *Discrete* __ *Memory* __

Form 4: Customer Test Flow Questionnaire

D. Product Power Supply:________________________

Power Rating: ______________________________

Max. Voltage Required : _______________________

Product's Current Consumption: ________________

III. TESTING REQUIREMENTS:

a. Max. Frequency: ______________________________

b. Test Channel Required: _______________________
Digital: ____________________________________
Analog: ____________________________________

c. Vector Memory Size: __________________________

d. No. of Power Supplies: ________________________

e. Any Option Required? Yes _______ No _______

e.1. Scan: Memory Size: ________________________
Max. Scan Chain: ____________________________

e.2. Analog:

- A/D, Conversion Speed: _____ No. of Bits: ______
- DAC, Conversion Speed: _____ No. of Bits: ______
- Others: ____________________________________

Form 4: Customer Test Flow Questionnaire (cont.)

e.3. Embedded Memory _____ Memory Product _____

- Memory Type: ______________________________
- Memory Size: ______________________________
- Address Pin: ____________ I/O Pin: ___________
- Read/Write Control ________________________
- Redundancy: ______________________________
- Test Pattern Required (BIST,...): ______________
- Others: __________________________________

f. Test Head Configuration: Please attach a copy of the configuration required.

g. What kind of test platform is required?__________

(Current Tester in Customer Site: _________________)

h. Test Service Required Items: Final Test___ Burn In___

*** *CUSTOMER TO PROVIDE TESTING FLOW* ***

IV. WAFER SORT REQUIREMENTS:

A. Test Platform Selection: ______________________

O/S Revision #: ____________________________

Prober Selection: ___________________________

Model #: __________________________________

Form 4: Customer Test Flow Questionnaire (cont.)

B. Test Program Revision: ____________________

B.1. Test Program: ____________________

B.2. Probe Card: Consign:___ Prepare by Vendor:___
Pin List / Diagram / Component Definition / Global File

B.3. Wafer Correlation: Consign ___
Apply Production Wafer ___

- Control Map____________________

- Correlation Data ____________________

B.4. Correlation Method: Run Binning___
Run Datalog___Run Shmoo___

B.5. Wafer Sort Temperature (C): ____________
Wafer Size: ____________ (inches/mm)

B.6. Test Time Per Wafer:____________
Die Thickness: ____________ (mils/micron)

B.7. Die Quantity per Wafer:____________
Die Size = X: ______ Y: ______ (mils/ micron)

B.8. Apply: Inking: __ Wafer Map for Defect Die: __
Ink-less: __

B.9. Low-Yield Setting: ____________________

REMARKS: ____________________

____

Form 4: Customer Test Flow Questionnaire (cont.)

V. FINAL TEST REQUIREMENTS:

A. Test Platform Selection: ______________________

O/S Revision #: ______________________

Test Handler Selection: ______________________

Model#: ______________________

B. Test Program Revision: ______________________

B.1. Test Program: CONSIGN

B.2. Load Board (Hi - Fix) / DUT Board:
Consign __ Prepared by Vendor __ Pin List - Diagram

Socket Type: ____________ Part #:____________
Contactor Type: ______________________

B.3. Correlation Samples: Pass Qty.: ____________
Fail Qty.: ____________

B.3.1 Correlation Method: Run Binning:_________
Run Data Log: ______ Run Shmoo: _________

B.3.2 Correlation Data: Consign: _____
Generate by Vendor: _____

B.4. Parallel Testing : YES / NO

If YES, number of Test Sites: ________

B.5. Testing Temperature (C): 1st: ______________
2nd: ______________ 3rd: ________________
Soak Time: __________ minutes

Form 4: Customer Test Flow Questionnaire (cont.)

B.6. AQL of QA Testing Gate: 1st: ___________
2nd: _________ 3rd: _________ Vendor std.: 0.10%

Kindly specify if tighter AQL required.

Is customer current practicing skip lot sample buy-off at electrical test? YES / NO

B.7. Estimated Testing Time: 1st: ________________

2nd: _______________ 3rd:_____________________

QA test time: 1st: ____________ 2nd: ____________

3rd: ______________

B.8. Low Yield Setting: 1st:______________________

2nd: ________________ 3rd: ___________________

Other Yield Control if any: _________________

B.9. Number of times required for 100% electrical test

(test insertion): ______________________________

B.10. Number of Good Bin: 1st: _________________

2nd: _________________ 3rd: __________________

REMARKS: ____________________________________

____

Form 4: Customer Test Flow Questionnaire (cont.)

VI. BURN-IN REQUIREMENTS:

A. Burn-In Selection: Dynamic Burn-In: ___________

Test During Burn-In (TDBI): __________________

A.1. Burn-In Board:
Consign: ____ Prepare by Vendor: ____ Both: ____

B/I Socket Qty per Board: ____________________

Pin List - Diagram - Wave Form - Voltage

- Special Request: ______________________________

- Matrix card: _________________________________

A.2. Burn-In Samples: Pass Qty.: _______________

A.3. Burn-In Condition: B/I Time: ______________

Temp. (C): _____________ Vcc: ______________

A.4. Post Burn-In Window Time: _______________

B. Auto Load / Unload Kit Preparation: __________

Burn-In Board: Qty. : _________________________

Preparation Time: _________________

C. Pre-Burn-In Test Required ? YES / NO

Specify Condition: ___________________________

REMARKS: ____________________________________

____

Form 4: Customer Test Flow Questionnaire (cont.)

VII. TOP-SIDE MARKING REQUIREMENTS:

A. Is Top-Side Marking Required? Yes ___ No ___

White Ink ________ Laser Mark ______________

A.1. Please Specify the Marking Requirements:

Size of Characters: ____________________________

Format: ______________________________________

Exact Location at Top Side: ___________________

**** Customer to provide a complete marking specification. ****

REMARKS: ____________________________________

VIII. LEAD SCANNING REQUIREMENTS:

A. Is the Lead Scan Required for:

__ Sampling QA Lead Scan Gate
(Std.: AQL: 0.10% - No Charge)

__ 100% Lead Scanning (Additional Charge)

NOTE: *Vendor guaranteed outgoing of 4.0mils coplanarity and 5.0mils bent leads at AQL 0.10%.*

Customer to specify limit if different from vendor control.

Criteria for: COPLANARITY: __________________
BENT LEAD: __________________

Form 4: Customer Test Flow Questionnaire (cont.)

B. If package is in trays, please specify the following (for lead scan and test handler setup):

Tray Vendor: __________ Bakable Temp: __________

Part#: ____________

Consigned ___ or Use Vendor Std. ___ or

Vendor Purchase on Behalf ____ or Vendor to

Purchase Based on Customer-Specified Tray Type: ___

REMARKS: ________________________________

IX. PACKING REQUIREMENTS:

A. Per Inner Box Quantity: ______

Required Combine ___ or Allow Vendor to Ship in

Partial Box for Remnant Lot ___

A1. Dry Pack Required? Yes ___ No___

If yes, please fill in criteria below:
Bake: ___ Vacuum Seal: ___ Both: ___
Baking Condition: ____ (C) ____hrs
Window Time: __________ hrs

NOTE : Vendor baking standard: 4 hrs @ 125C.
Window time: 12 hrs. (or 24 hrs if tape & reel required).
Normally, baking time is based on customer requirement.

Form 4: Customer Test Flow Questionnaire (cont.)

A2. Dry Pack Material: Dry Pack Bag : ____________

Vendor:__________ Bag Model #: _____________

Desiccant: 1-unit ____ 2-unit ____ 4-unit ____

No. of Desiccant Bags Required: _________

Humidity Indicator Card: 6-spot: ___ 3-spot: ___

A3. If Use Vendor Standard: _______

Dry Pack bag: Vendor: _____

Static Control Components Bag Model #: ________

Additional Information:
Ship Full Inner Box / Combine Lot ___ or
Allow to Ship Partial or Remnant Lot ___

REMARKS: ___________________________________

X. TAPE & REEL REQUIREMENTS:

A. Are Tape and Reel Required? Yes: ____ No: ____

If yes, Quantity/Reel: _________________________

Tape and Reel Material Supplier: _______________

Carrier Tape Part #: ___________________________

Cover Tape Part #: ____________________________

Lockreel Part #: ______________________________

Form 4: Customer Test Flow Questionnaire (cont.)

Vendor Standard Tape and Reel Material Supplier:

Advantek / PEAK

Carrier Tape Part #: Base on Package Type

REMARKS: ______________________________

XI. IC CONTAINER REQUIREMENTS:

A. Please Specify Required IC Container:

__Tube: Tube Size :______ Qty./Tube: ______
Tube/Inner Box : __________

__Tray: EIAJ :__ JEDEC: __ Supplier: _______

Temperature (C):__________ Part#: __________

Tray or Tube Qty./Inner Box: _______________

NOTE : Tube drawing (if it's a tube product).

REMARKS: ______________________________

Form 4: Customer Test Flow Questionnaire (cont.)

XII. INNER BOX REQUIREMENTS:

A. Is the Inner Box: __ Use Vendor Std. __ Vendor to Prepare for Customer (with Customer Logo)

NOTE : *Prefer customer to provide a copy of the packing specification/requirement with the following information:*
* *customer box drawing with graphic (inner and outer shipping box)*
* *packing barcode label requirements*
* *attached documents to packing*

NOTE: *Additional tooling cost might be applied if quantity is small.*

REMARKS: ________________________________

XIII. OUTER BOX REQUIREMENTS:

A. Is the Inner Box: __ Use Vendor Std. __ Vendor to Prepare for Customer (with Customer Logo)

NOTE : *Prefer customer to provide a copy of the packing specification/requirement with the following information:*
* *customer box drawing with graphic (inner and outer shipping box)*
* *packing barcode label requirements*
* *attached documents to packing*

Form 4: Customer Test Flow Questionnaire (cont.)

NOTE: Additional tooling cost might be applied if quantity is small.

REMARKS: ______________________________

XIV. PACKING LABEL REQUIREMENTS:

A. Is the Packing Label: __ Use Vendor Std.

___Vendor Test Prepare for Customer:

Dimension: ____________ Samples: ___________

NOTE : Customer to provide a sample for Vendor to generate label with matched font size, and also to work out label automation to prevent manual entry by operators.

REMARKS: ______________________________

XV. SHIPPING REQUIREMENTS:

A. Shipping Requirement:
__ Normal Shipment (Ship Back to Customer)
__ Drop Shipment (Directly Ship to End Customer)

A.1. Please Specify all Shipping Documents Required:

____

REMARKS: ______________________________

Form 4: Customer Test Flow Questionnaire (cont.)

XVI. CUSTOMER REJECT PARTS HANDLING REQUIREMENTS:

A. Are the Customer Reject Parts Authorized to Be: (select one)

__ Shipped out together with the good parts and with proper identification to customer ?

__ Scrapped Automatically at Vendor?

__ Shipped Back to Customer on an Agreed Period?

__ Other Option? Please Specify:______________

REMARKS: ________________________________

Form 4: Customer Test Flow Questionnaire (cont.)

XVII. ADDITIONAL REQUIREMENTS:

A. List of specifications to be provided by customers:

___ a. Process Flowchart
___ b. Test correlation procedure
___ c. Product qualification requirement
___ d. Operational procedure
___ e. Packing/Bake requirements (attach packing sample)
___ f. Visual mechanical criteria
___ g. Package drawings
___ h. Tray & tube drawings
___ i. Inner & Outer boxes drawings (attach box sample)

XVIII. OTHER REQUIREMENTS:

XIX. CUSTOMER COMMENTS :

Form 4: Customer Test Flow Questionnaire (cont.)

THE COT PERSPECTIVE

Primer on Assembly and Packaging of Semiconductors

Satya Chillara
Director Corporate Marketing
ChipPAC, Inc.

Assembly and packaging has become a highly critical step in the manufacture of semiconductors. No longer an afterthought, today's packaging must be factored into the process early in the design cycle. Starting with a metal can and DIP packaging, the technology has now evolved into assembling complex packages up to and including ~2000-pin flip-chip BGAs. According to Dataquest, there are more than 1000 different types of packaging, which creates a significant challenge for operations and packaging people in both fabless and IDM companies to determine the right packaging solution for each given device technology.

The different types of packaging within the industry are broadly categorized as leadframe or substrate. Leadframe is predominantly used to package 2- to 250-pincount devices in a variety of styles, including plastic dual-in-line package (PDIP), PLCC, SO, QFP and TO. Substrate technology typically is used for 8- to 2000-pin packages in the form of Ball Grid Array (BGA) or Chip-Scale Packaging (CSP). In terms of interconnection, wirebond, TAB and flip chip methods are all used, with more than 95 percent of the volume carried out in wirebond.

Leadframe Packaging

Leadframe or leaded packaging has been in existence since semiconductors were first produced. It is a mature technology and accounts for more than 95 percent of the current volume of semiconductor packaging. Leaded packaging is found in virtually every electronic product, including automobiles, household appliances, desktop and notebook computers, and telecommunications devices. The major styles of leadframe packaging are SO and QFP.

Leaded packages are characterized by a semiconductor die, which is attached to a leadframe or flat lattice of wires and then encapsulated in a plastic or ceramic package. The ends of the leadframe wires protrude from the edges of the package to enable connection to a printed circuit board. Originally designed to be plugged into the PC board (plated through holes), this technology has evolved into a more modern, surface-mount design in which the leads or pins are soldered to the surface of the board. Specific package customization and improvements are continuously being engineered to improve electrical and thermal performance, to shrink package sizes, and to enable multi-chip capability in leaded packaging.

Leaded Package Profile

In leaded packaging, the industry offers a variety of lead counts and body sizes to satisfy customer die-size variations. Standard packages are at least three millimeters thick and include PDIP, PLCC, and SOIC varieties; while the more advanced leaded packages are thinner, approximately 1.4 millimeters thick. Advanced leaded packages have a finer pitch because the leads are closer together, which allows for a higher pin count and greater functionality in a smaller package size. Advanced leaded packages include MQFP, TQFP, TSSOP, thermal QFP, leadless CSP, and SSOP.

Substrate Packaging

Substrate packaging, such as BGA or CSP, represents the newest and fastest-growing segment of the assembly and packaging industry. Substrate packaging is used primarily in high-growth end-markets such as computing platforms and networks; hand-held consumer products including wireless communications, personal digital assistants, and video cameras; and home electronic devices such as DVDs and home video game machines.

BGA technology was developed to solve the problems associated with the increase in lead counts required for advanced semiconductors. As the number of leads surrounding the IC went up, significant electrical shorting problems occurred. BGA solved this problem by effectively creating leads on the bottom surface of the package in the form of small bumps or solder balls. In a typical BGA package, the semiconductor die is placed on top of a plastic or tape laminate substrate rather than on a leadframe. The die is connected to the substrate by a series of fine gold wires that are bonded to the top of the substrate near its edges. On the bottom of the substrate is a grid of metal balls that connect the packaged device to the printed circuit board.

The benefits of BGA packaging over leaded packaging include

- Smaller size,
- Higher pin count or number of connections to the PC board,
- Greater reliability,
- Better electrical signal integrity, and
- Easier attachment to the printed circuit board.

BGA Package Profile

BGA and CSP packages are available in a number of styles to address electrical, thermal, and thinness needs.

Standard BGA

Standard BGA packaging has a grid array of balls on the underside of the integrated circuit and is used in high-performance applications like personal computer chipsets, graphic controllers, and DSPs. A standard BGA package generally has more than 100 pins and provides better thermal and electrical performance than leaded packages. BGA also features a more advanced surface mounting technology, which provides easier handling during the packaging process.

Thermal and Electrical BGA

Thermal BGA comes in "cavity-down" and "cavity-up" versions. Cavity-down thermal BGA can be obtained in multiple layers (up to six) to enhance electrical performance. Cavity-up thermal BGA incorporates a drop-in heat slug and can also be produced in six layers or more to enhance electrical performance. Thermally and electrically enhanced BGAs are available in both laminate and tape substrates, with today's tape substrates restricted to versions with two metal layer.

Flip-Chip BGA

In flip-chip BGA, silicon die is attached directly to the substrate using bumps or solder balls rather than wire bonds. This is expected to improve the heat dissipation as well as the electrical performance of the die. Flip-chip BGA can be used in a wide number of applications ranging from consumer products to highly sophisticated, application-specific integrated circuits (ASIC), microprocessors, and memory packages. Flip-chip BGAs are available in either ceramic or organic build-up as well as in PTFE-based substrates.

Chip-Scale Packaging

Chip-scale BGA is a substrate-based package, where the size of package is 1.2 times the size of the actual silicon die. It is typically designed for memory devices and other medium-pin-count semiconductors. It requires dense ball arrays in very small package sizes like those used in wireless telephones, personal digital assistants (PDAs), video cameras, digital cameras, and pagers.

Table 6 below provides a snapshot of the typical pin counts and applications associated with various packaging types.

Option	Types	Applications	Pin Count
Traditional Leaded Packaging	PDIP, PLCC, SOC, SOJ, SSOP, SIP, DPAK, D²PAK and TO	Telecommunications, automobiles, household appliances, desktop and notebook computers	8-64
Leaded Packaging	TSOP, TSSOP, MQFP, LQFP, TQFP, and thermal versions of leaded packages	Personal computers and telecommunications	32-304
Standard BGA Packaging	PBGA, Multi-chip BGA cavity down and drop-in heat slug cavity up thermal BGA	Personal computer chipsets, ASICs, and graphic controllers	119-900
Chip-Scale BGA Packaging	Single, stacked multi-chip CSP in laminate and tape formats	Wireless phones, PDAs, digital cameras, and camcorders	8-324
Flip-Chip BGA Packaging	Flip-chip BGA, wafer level CSP and flip-chip CSP	High-end network servers, routers and microprocessors, and memory packages	2-2000

Table 6: A Wide Variety of Packaging Options and Their Possible Application Types and Pin Count Ranges

ChipPAC, Inc.

Table 7 shows the relationship between fab geometry and package interconnects. At 0.18 microns, most customers were implementing 40/80-micron staggered, 60-micron in-line wirebonding, and 225-micron bump pitch flip chip. As geometries shrink to 0.1 microns, wirebonding pitch is expected to come down to 30/60-micron staggered, along with 40-micron in-line wirebonding and 180-micron bump pitch flip-chip.

0.09um					**30/60um Staggered 40um In-line 180um Flipchip**
0.13um				**35/70um Staggered 50um In-line 200um Flipchip**	
0.18um			**40/90um Staggered 60um In-line 225um Flipchip**		
0.25um		**45/90um Staggered 70um In-line 225um Flipchip**			
0.35um	**60/120um Staggered 120um In-line**				

Table 7: Relationship between Fab Geometry and Package Interconnects

Packaging Design, Simulation, and Characterization

Packages are designed using various tools such as the Cadence APD (Advanced Package Designer), Avanti's Encore, and the AutoCAD system. The key requirements from the customer are package I/O and electrical, mechanical, and thermal specifications. Once these requirements are understood, the design team can design, simulate, or characterize the package to meet all customer requirements. The electrical analysis of the packages is done using tools such as Ansoft to obtain inductance, resistance and capacitance (LRC) values. Also, detailed system-level electrical simulation and characterization analysis can be carried out. Thermal analysis provides thetaJA and thetaJC values. Detailed system-level thermal simulation and characterization can be carried out. Mechanical analysis provides insights into board-level temperature, power, and bend cycling to understand the interaction between die, package, and PC board.

Package Interconnect

The interconnect mechanisms being practiced in package assembly today are wirebonding, tape automated bonding, and flip chip, with the vast majority being done with wirebond. Flip chip is practiced for performance-related applications such as microprocessors, ASICs, and chipsets.

Within wirebonding, either in-line or staggered wirebonding can be used. Today, in-line wirebonding is being carried out at 50-micron pad pitch and is expected to go down to 40 microns as fab geometries shrink to the 0.1-micron milestone. With respect to staggered wirebonding, 40/80-micron pad pitch bonding is practiced and expected to shrink to 30/60-micron as the fab geometries reach 0.1 micron. Also in the staggered bonding arena, customers are evaluating three-row staggered wirebonding to save the die area. Figure 13 shows typical cross section photographs of in-line, staggered wirebond, and flip-chip interconnects.

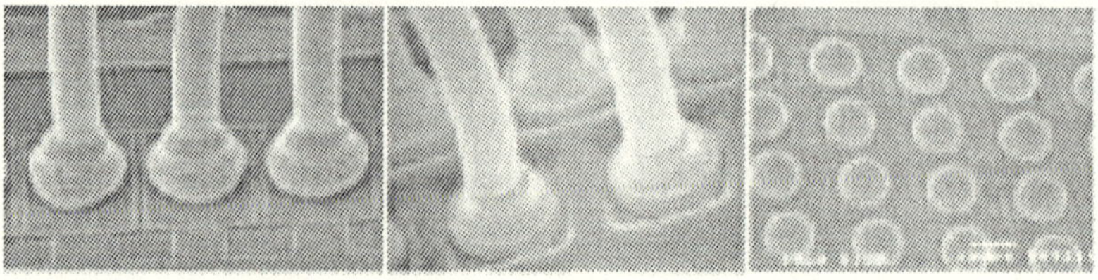

45um Pad Pitch 35/70 Staggered Pitch Flip-Chip Bump Pitch

Figure 13: Cross-Section Photographs of Wirebond, Staggered, and Flip-Chip Interconnects

Substrate technology, too, has evolved to keep pace with the advances in wirebonding. In the table below, wirebond design rules on multiple substrate technologies (laminate or tape/build-up(BU)/thin film) are given. With a die pad pitch of 60 microns, laminate substrates need to maintain a finger pitch of 100 microns using a 25-micron wire diameter. As seen in Figure 14, more advanced substrates, such as tape, build-up or thin film configurations, are needed to reduce the line and space width of the substrates to keep up with reduction in wire diameter. Figure 15 shows the stitch wire on the 135- and 125-micron finger pitch substrates.

Substrate	Laminate					Tape/BU/Thin Film					
Finger Pitch (um)	145	135	125	115	100	90	80	75	55	45	35
Top Width (um)	75	70	65	60	50	40	35	30	25	20	18
Min Space (um)	35	35	35	30	30	30	30	30	25	20	15
Trace Pitch (um)	120	110	100	90	80	60	60	60	50	40	35
Wire Dia. (mil)	30	30	28	28	25	20	18	15	13	10	9

Figure 14: Design Rules for Die, Trace, Space, and Finger Pitch of Substrates

The COT Perspective

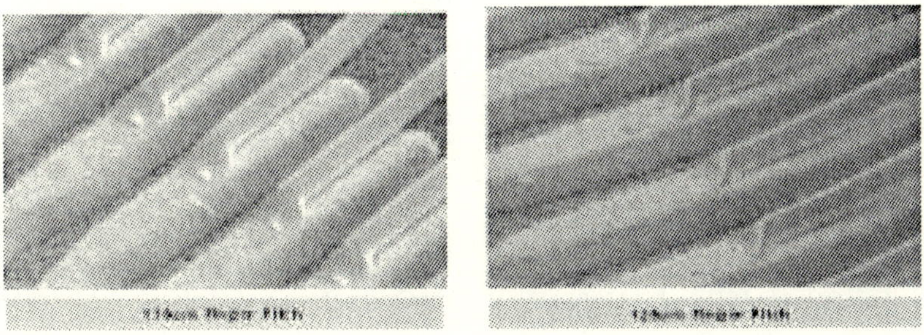

Figure 15: Cross-Section of Stitch Wire on the 135- and 125-Micron Finger Pitch Substrates

Assembly Flow

Assembly begins when the foundry completes fabrication of the IC circuitry on the wafer. Wafers are then back-grinded to a particular thickness depending on the package form factor. For example, wafers are back-grinded to 11 mils in the case of a 1.2 mm TSOP, and to 5.5 mils in the case of a stacked CSP. After back grinding, wafers are mounted on a tape and diced, then inspected for contamination or foreign particles. Meanwhile, substrates are pre-baked to get rid of moisture and the individual die is attached to the substrate with a die-attach paste which has been cured at 175 C for 5 to 15 minutes. The next step is to carry out plasma operations on the die (which is sitting on the substrate) to get rid of any contamination on the die pads and ensure better wirebonding. The wirebonding step is carried out at approximately 200 C. Then a third optical inspection is performed to make sure there are no lifted wires. Again, plasma operations are carried out for better mold adhesion to the solder mask of the substrate, and the post-mold cure (PMC) is completed to cure the mold compound. After PMC, the packages are laser-marked per customer marking instructions.

Solder balls are mounted on the back of the substrate, and they are sawed into individual packages. After sawing, the packages are scanned for co-planarity of the balls and an external visual inspection (EVI) is performed to ensure that the package has no missing balls. The final step is to put the packages in a tray and ship them per the customer's instructions.

Package Reliability and Lead-Free Packaging

Once assembled, packages are subjected to a variety of reliability tests. There are a number of variables in a package (die thickness, substrate types, package thickness, ball pitch, and ball diameter), which ultimately result in a different package form factor. For example, a 0.5mm-thick die is usually used in a plastic, dual-in-line package (PDIP), while a 0.14 mm thick die is used in a stacked CSP version (M2CSP-SDS). Similarly, different substrate materials are used to achieve overall thickness in a given package. The thickness of the substrate as well as the overall package thickness plays a major role in the board-level reliability of the completed device. Also, solder pitch and diameter play a major role in board-level reliability (solder joint reliability gets better with larger solder ball pitch and ball diameter).

Today, lead-free packaging is also gaining in importance as Japan, Europe, and the United States work toward removing lead from electronics packaging. In removing lead, surface mount flow temperatures increase from 240 to 260 C. Different bills of materials have been developed to withstand these temperatures and to meet Joint Electron Device Engineering Council (JEDEC) moisture-level conditions.

The COT Perspective

Board-Level Reliability

Board-level reliability (BLR) has become a standard in the packaging industry with ever-growing miniaturization. As packages become smaller and smaller to fit within the newer, smaller-end products, BLR becomes a major concern. BLR methodology is started with the modeling of key characteristics and the gathering of experimental data for validation. Once sufficient data has been assembled and the model fully validated, simulation alone can be used instead of actual BLR tests which are time-consuming and costly. Periodic verification of BLR tests is performed to continuously re-validate the model. Every application, including cell phones, camcorders, and PCs, dictates a certain temperature cycle performance, as well as related requirements such as bend testing, drop tests, and power cycling.

Package Costs

Each semiconductor package type is designed to meet certain end-user or applications requirements. For example, a 208-pin package can be offered in a 28 x 28 mm, 0.5 mm pitch, QFP package or in a 15 x 15 mm CSP package at 0.8 mm ball pitch. So, the same pin count can be in different form factors and be available at different price points. The customer's product defines what package form factor is the most appropriate type. Typical examples would be a 208 QFP package used in a DVD layer and a 208 CSP used in a cell phone.

ChipPAC, Inc.

Assembly Request Form

On the following pages is a typical assembly request form that packaging companies expect their customers to fill out, giving an opportunity for both the customer and the supplier to document all the requirements for correctly designing and assembling the product.

The COT Perspective

Typical Assembly Request Form

I. Original Request Date

Year (YYYY): 2001
Month (MM): 2
Day (DD): 5

II. Contact Information - Customer

Company Name: Customer
Project or Device: Project
Application: Wireless
Name: John Customer
City/State/Country: Sacramento/CA/USA
Voice Number: 555-555-8777
Fax Number: 555-555-8771
Email Address: john.customer@customer.com

III. Contact Information – Sales Representative

Name: John Sales
Voice Number: 000.000.0000
Fax Number: 000.000.0000
Email Address: john.sales@chippac.com

IV. Request (Select all that apply)
O IS A QUALIFICATION BUILD DESIGN

O	Design Feasibility	(X) Characterization – Thermal
(X)	Full Design	- AmbientTemp (C): 24
O	Design – Rule Check	- Power (W): Identify in Section VI
O	Design Documentation	(X) Characterization – Electrical
		- Oper. Freq. (MHz): Identify in Section VI
		(X) Characterization– Board Level Thermal Cycle Reliability
		- Max. Temp (C): 125-Ramp Time(Min): 15
		- Min. Temp (C):-40- Dwell Time (Min): 15

Other Comments (Please indicate Package Qualifications Requirements):

Request completion by 02/25/02.

Form 5: Typical Assembly Request Form

V. Product Family

XYZ Body Size (mm): 35x35x1.4		Lead Count:	352
Layer Count:	1	Lead Pitch (mm):	1.27
Substrate Thickness (mm):0.61		Lead Standoff (mm):	0.76
		Ball Diameter (mm):	0.35

O PBGA	O FBGA-T	O iQuad (1)	O Other Package
O TBGA-I	O uBGA	O iQuad (4)	
(X) TBGFA-II	O EconoCSP-L	O QFP	
O TBGA-III	O EconoCSP-T	O TSOP	
O M^{2}BGA	O LFCSP	O TSSOP	
O FlipPAC	O M^{2}CSP	O PDIP	
O LGA	O M^{2}CSP Same Die Stack	O SOIC	
	O Flip Chip – CSP	O SSOP	
	O WLCSP	O TO (2XX)	

XY Cavity size for cavity – down packages (mm): 11.1x11.1

Other Comments:

No Other comments

VI. Device Information

Device #1	Project		
XYZ Size (mm):	10x10x0.3	Device #3	
I/O Count:	380	XYZ Size (mm):	
Wire Dia. (mils):	1.0	I/O Count:	
Power (W)	2.5	Wire Dia. (mils):	
Oper. Freq. (MHz):	125	Power (W)	
XY Pad Pass. Open (um):	65x65	Oper. Freq. (MHz):	
		XY Pad Pass. Open (um):	
Device #1			
XYZ Size (mm):		Device #4	
I/O Count:		XYZ Size (mm):	
Wire Dia. (mils):		I/O Count:	
Power (W)		Wire Dia. (mils):	
Oper. Freq. (MHz):		Power (W)	
XY Pad Pass. Open (um):		Oper. Freq. (MHz):	
		XY Pad Pass. Open (um):	

Form 5: Typical Assembly Request Form (cont.)

The COT Perspective

Other

No other comments

VII. Assembly Information for wire-bonded packages

Die Attach Material: (X) Conductive Epoxy O Non-Conductive Epoxy O Soft Solder

Wire Diameter: O 1.2 mil O 1.1mil (X) 1.0 mil O 0.8 mil O Other

Molding Compound:(X) Encapsulation O Transfer Molding O Other

Solder Finish (L/F) O Tin/Lead (X) Lead-Free

Solder Balls (Substrate) O Tin/Lead (X) Lead-Free

Shipping (X) Tray O Tube O Other

VIII. **Assembly Information for Flip Chip Packages**

Eutectic Solder Bump O Electroplated O Screen Printed O Other

Heat Spreader O Yes O No

Solder Balls (Substrate) O Tin/Lead O Lead-Free

IX. Volume Estimates (Include units of K or KK) and Calendar Information

Current Year:10K Customer Die Tape-out (MM/DD/YY): 03/15/02
Current Year + 1:20KKReq'd Substrate Receipt (MM/DD/YY): 03/16/02
Current Year + 2:20KKReq'd Assembly Shipout (MM/DD/YY): 03/20/02
Current Year + 3: 5KK
Current Year + 4: 1KK

Form 5: Typical Assembly Request Form (cont.)

End Product Application: Cellular Handset

Life Time of Product: 5 years

XI. Device Netlist

Device Name	Net Name	Die Pad No.	Pad X-Coord(um)	Pad Y-Coord(um)	Pkg. Pin No.
D1	NNI	1	1000	1025	A1
D2	NN2	2	1040	1025 .	A2

Pad XY-Coordinates should be to the center of the XY pad passivation opening (as identified in section VI – Device Information) with the device name center as the origin.

Past Tab Delimited text File Netlist

Paste text based netlist here

Form 5: Typical Assembly Request Form (cont.)

The COT Perspective

Major Points of Chapter 5

- The typical CMOS manufacturing flow includes wafer fabrication, wafer probing, packaging, assembly, testing, and delivery.

- It is easier from a logistical point of view to work with one assembly house for all your back-end requirements.

- Choosing the right assembly house or set of vendors should be based on capability, geography, pricing, cycle-time, capacity, and relationship. The most important of these is capability.

- It is up to you to verify your vendor's capabilities.

- The best way to assure you are getting the best price for service is to obtain competing quotes from multiple vendors.

- Don't be afraid to challenge the assembly house with questions about support and cycle time.

- Prior to submitting your design, ensure that your design does not violate assembly house design rules and cause bonding violations such as angles, crossovers, or bond pad pitch violations.

- Depending on the package type, you may need to order package materials to supply to your assembly house.

- Good planning and an attitude of never assuming anything will help you avoid catastrophes.

- The goal of package and test planning is to minimize costs in production.

- Ten years ago, testing represented less than 10 percent of the total cost of manufacturing a commercially-used IC. Today this percentage can approach 40 percent.

- Design your test requirements to use the least expensive ATE.

- Limit test time through ATE.

- Subcontractors can help reduce your overhead, while maintaining time-to-market requirements.

- The increase in package costs is making probe tests much more important.

CHAPTER 6

Design Tapeout

In the last chapter, we discussed how to select and engage a foundry and back-end vendors necessary for your chip manufacturing. This chapter focuses on the details of the tapeout process with your foundry.

After locking in with a foundry, keep in mind that a great deal of planning and anticipation of requirements needs to happen in order to ensure a smooth experience through fabrication. Let's start by investigating the first step in the tapeout process flow.

The three major phases of the tapeout process flow are highlighted in figure 1 below. Phase 1 is the data prep and check phase, phase 2 is the actual act of taping out, and phase 3 includes mask tooling and the manufacturing start.

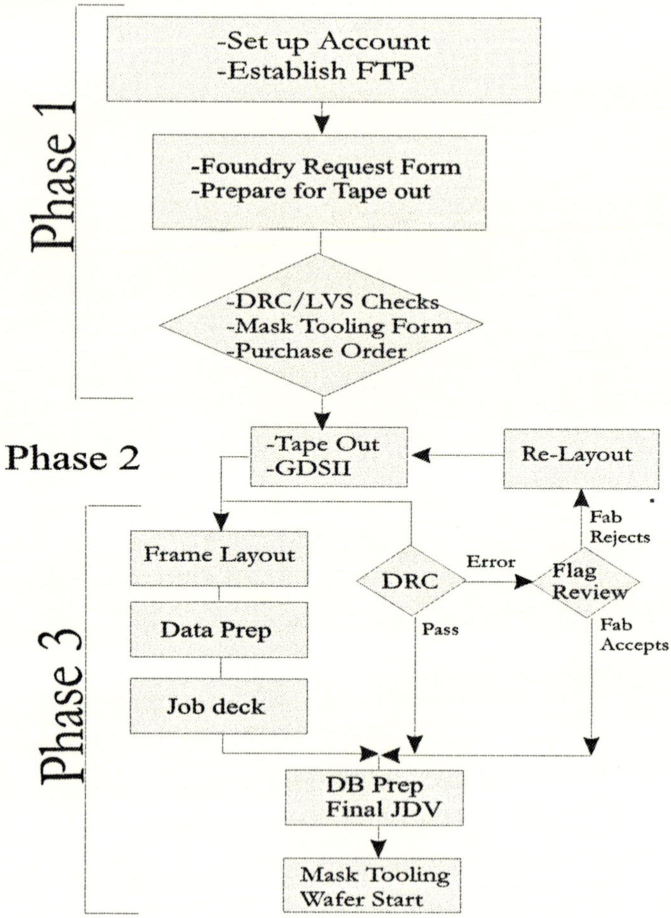

Figure 17: Typical Tapeout Process Flow

Phase One: Pre-Tapeout

The pre-tapeout phase involves setting up an account with the foundry and ensuring that your design database is ready to send to the factory.

Setting Up an Account

After you have negotiated prices and agreed on a basic contract, you need to set up an account with the foundry. Setting up an account normally includes establishing credit and setting up an FTP account. (The FTP account is where you will be sending your completed design database.) Examples of typical forms that the foundry will ask you to fill out are provided at the end of this chapter.

Establishing Credit

The next step is to establish some credit with your foundry. Given that foundry services such as mask tooling and wafer generation are very expensive, and given that these costs are increasing, it is understandable that the foundry wants to ensure that your company is creditworthy before building up an expensive mask set. (At the time of this writing, typical mask sets for 0.15um cost more than $500K, and these prices are expected to double for 0.1um.)

In the process of establishing credit, the foundry will ask to speak to your bank. Foundries will typically obtain your approval to approach your company's bank by having you fill out and sign an authorization form. A foundry may also send your bank a questionnaire about your credit worthiness.

Once the foundry has established that your company is creditworthy, you can move to the next step.

Setting up an FTP Account

Most foundries today provide you with an FTP site that is password-protected and is built with a firewall to protect your database. The FTP application is straightforward. It is a good idea to ensure you have set it up in advance of your tape out. The foundry will also supply FTP procedures for how to download your files into the protected site. You may want to encrypt your database before transferring it to an FTP site and provide the foundry with the key to unlock it.

The actual setting up of the FTP account, of course, is handled by the foundry. It is best to check with your foundry to find out which FTP procedures they want you to follow.

Foundry Service Request Form

The foundry service request form serves as a preliminary "heads up" to the factory. This form outlines to the foundry the general services that you are expecting. (Please refer to the example form provided at the end of this chapter.)

As you can see from the example, the foundry request form simply outlines the basics of the technology flow for your design. It is also a higher-level check to make sure the foundry knows what to do with your design once wafers are available. Typically, there are questions in the form about wafer sort, assembly, and turnkey. Basically, the foundry wants to make certain that you have thought out how you will have your die sorted and packaged. In a turnkey scenario, the foundry performs these services. (As discussed previously, turnkey is not normally offered by foundries without a commitment for high-volume business. Smaller companies usually need to make separate arrangements.)

Design Rule Compliance

Of critical importance prior to tapeout is to ensure that your design meets your foundry's latest design rule versions and SPICE models. A check to ensure that you pass LVS is also necessary prior to tapeout.

Do not assume that the foundry will perform these checks. They may have a policy to check DRCs for new customers, but it is not guaranteed.

Keep in mind that design rules often change or get updated. This is especially true for cutting-edge process technologies. In fact, if you have decided to use the newest process technology available, you can expect the design rules and SPICE models to change frequently. This is due to the fact that foundries are constantly optimizing performance and production efficiencies, especially for the new processes. Of course, the newer the process, the greater the need for adjustments that affect the design rules and SPICE models. As a designer, you should be constantly checking with the foundry to make certain your models and design rules are of the latest revision.

What applies to you also applies to third-party IP in your design. The IP blocks must be compliant with the latest revisions as well. Don't assume that the IP vendor is aware of the latest foundry revisions. Make certain by asking the foundry exactly what DRC revisions they are compliant to.

As discussed earlier, companies going through the manufacturing experience for the first time often forget this point about third-party IP. It happens quite frequently that designs reach the foundry with incorrect DRCs. Be vigilant and maintain a close working relationship with foundries as well as with your IP vendors. In an ideal world, the foundry and your IP vendors should warn you if design rules are changing. But the world is far from ideal. Design rules can change right before designs arrive in the factory. Avoid getting surprised by keeping the foundry's salesperson in close contact as you go through tapeout.

Generally, you need to worry less about big-name IP vendors with an established customer base and working designs. They are most likely working closely with the major foundries to prove their IP with the latest processes. Also, foundries will often publish lists of IP vendors who have performed various levels of compliance tests at their respective foundries. This is a good guide, but it is still important to ensure that your particular block has been compliance tested. Even with the big IP vendors, however, it is still a good idea to double check. Unfortunately, since the whole industry is resource-constrained, surprises are common.

IP Usage Form

Part of the pre-tapeout process includes filling out a form that lets the foundry know what IP you are using in your design. The foundry *does not* use this form to double-check IP compliance. This form is primarily used for royalty payment tracking. As we explained earlier, foundries often offer customers certain design libraries from third-party vendors free of charge. While it may be free to you, in fact the foundry is picking up the bill. Some foundries track and manage these royalty payments using the IP usage form. Today foundries may also use software that automatically tracks the use of their IP blocks. In any event, keep in mind that the IP usage form does not eliminate the need to make certain that third-party IP is compliant to a given process technology.

Mask Tooling Form

As the tapeout date approaches, you will need to prepare the mask tooling form. An example form is provided below.

Design Parameters and Requirements

The major purpose of the mask tooling form is to confirm your design parameters and requirements. The form highlights any new changes to your design strategy or manufacturing requirements that may have occurred since the time you first approached the foundry with your project plans. For example, your designers may find that certain logic performances require extra mask layers, or need passive L and C components for better performances of some analog functionality. The foundry wants to make doubly sure that the process flow that you originally requested before tapeout is still the same at tapeout. The foundry also uses the mask tooling form to cross-correlate the names and references of mask layers to the names and references that happen to be used by your CAD program.

High-Level Design Issues

In addition, the mask tooling form clarifies basic high-level design issues such accommodating the space of the die seal ring. (If you are a start-up, it is advisable to allow the foundry to specify the die seal ring, because they will be more familiar than you are with a certain process technology and its foibles. Unless you have a lot of experience in this area and are familiar with the foundry and the process, letting the foundry make this call is your best option.)

Another high-level design issue is the treatment of your database. For example, the foundry will insert test keys into the scribe lines for WAT (wafer acceptance test) monitoring. As you will see in question 14 of our mask tooling form below, the foundry can offer you the opportunity to put in your own keys in case you have a specific test of your own. It is unlikely that you will have to worry about this as a start-up, but in some instances you may have special requirements. For example, if your devices implement high-performance mixed-signal elements that you expect are process control module (PCM) sensitive, it may be wise to create your own test keys to monitor certain criteria.

WAT Parameters

The mask tooling form will also confirm what set of WAT parameters the design will be manufactured to. Typically, the foundry's standard WAT parameters should be sufficient for your design. A company that might work with a foundry to use a special set of WAT parameters is more likely to be an existing IDM that has a long-running design with well-known performance sensitivities.

Wafer Handling Instructions

Question 15 of the example mask tooling form below allows you to specify wafer handling instructions. Most proto-lots are handled as hot lots by default; you may be able to negotiate a super-hot lot status, which could shave many days of cycle-time on your average design. In addition, a typical proto start is 12 wafers. The good news is that foundries will normally shoot an extra wafer for your probe card development at no charge. But you need to remember to request this service. The foundry will also typically ask about other wafer handling issues which are not normally required for first-run silicon, such as backlapping[1] and epi material insertion. Check these only if you have a clear requirement for these procedures. Likewise, it is unlikely you need corner splits on a new design and proto-run. Typically, the foundry will not even support corner splits on a proto-run since functionality and time to market are more important issues at the proto stage.

Purchase Order

The form is also used to make doubly sure that you have sent the foundry a purchase order (PO). Imagine a foundry's embarrassment if $500K worth of mask tooling is developed without a PO. Foundries are getting strict about this issue and will likely require a PO prior to a tapeout.

Database Cross-Correlation and Size

Page three of the example mask tooling form below is used to help the foundry cross-correlate the design database and ensure that all the data is available to the mask tooling shop. The foundry will often ask you to correlate the names of the references that your CAD files use to describe your design layers to

1 Backlapping is a process of grinding the back side of a wafer to make it thinner. This thinning process helps to eliminate stacking faults by creating stronger molecular bonds in the substrate material. Similar to etching, the process is used to minimize quiescent and leakage currents in gate oxide. For larger wafers, backlapping is not recommended until after wafer probe since the backlapping process reduces the structural strength of the wafer, making it too weak to be probed.

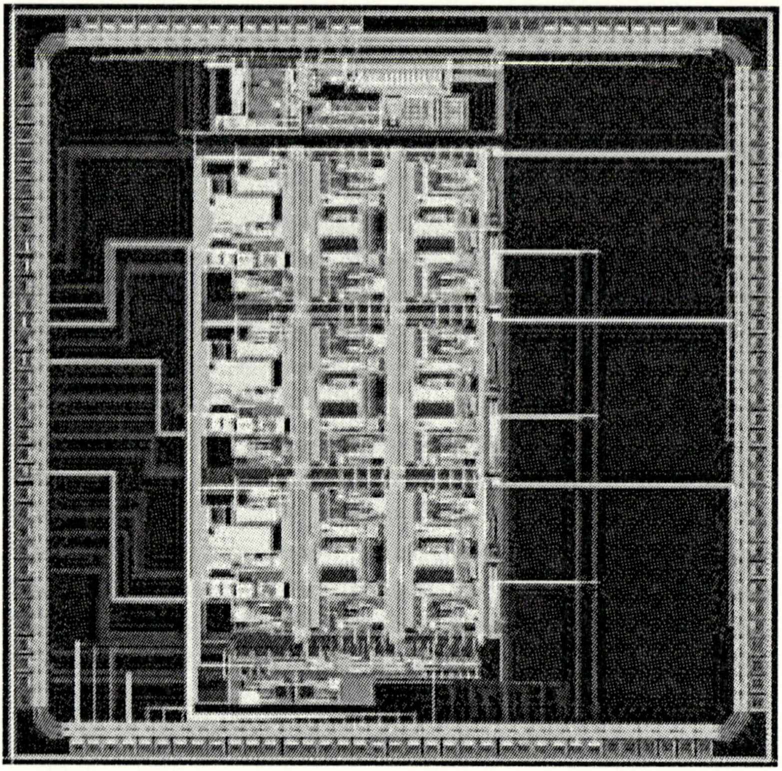

Figure 17: Chip Layout (Courtesy of UMC)

the standard names and reference numbers used by the foundry. The initial quote normally specifies these references. This step helps to minimize mistakes due to misinterpretations of your CAD files.

The mask tooling form will also ask you to specify the size of your database. It is common for data to be lost or corrupted during the FTP process. Therefore, it is always wise to ensure that the sizes of the design file sent and the file received by the foundry are the same.

The mask tooling form should be sent out a few days before tapeout to allow time to catch any problems or issues before mask tooling gets underway. Don't be concerned if your design has changed from the time you initially arranged your account with the foundry. However, keep in mind that certain changes such as extra masks or special wafer handling procedures have cost implications.

Phase Two: Tapeout

Tapeout is the act of sending your database to the foundry. If you are a new customer with a new design, the foundry may run a DRC upon receiving your database. (Typically, if you are a start-up, you will be submitting GDSII data. The foundry can do a DRC on the GDSII file, but they cannot at the time of this writing run DRCs on MEBES data.)

It is a useful trick to run DRCs on the database after FTP anyway. This is to ensure that your database was not corrupted during the transfer. Inexpensive double-checks like this can prevent major headaches and expenses later.

If your design does not meet the foundry's design rules, 99 percent of the time you can expect the foundry to request that you alter your design to meet the particular rule you violate. However, depending on the violation and the degree of variance from the norm, you may be able to negotiate with the foundry.

Foundries hate to do this, and in most cases it is not advisable. We just mention it here because you should be aware of the possibilities. Sometimes, if you are time-constrained you may want to run a prototype lot quickly and you may find yourself negotiating with the foundry to let them pass one lot through with a design violation with the promise that you will correct the violation in production. Some design violations may not actually have an effect on the design performance but rather may affect yields. Once again, it is advisable to avoid design violations altogether.

Most likely you will be sending your design to the foundry in GDSII format for tapeout. Most foundries will also accept MEBES files, which are actually mask tooling data. But some companies will want to control the mask tooling (normally more established IDMs). We mention this here only as an FYI; otherwise this topic is beyond the scope of this book.

Be aware that the price you pay for your silicon is always subject to change. If you make some radical changes to your design process flow requirements, you may pay more or less depending on how your requirements have changed. As a simple example, if you originally thought you would use four layers of metal but actually end up with five, you can expect to pay a higher charge per wafer. Of course, the opposite can happen as well, if you start out thinking you need a more exotic process flow and end up with cheaper requirements. In general, however, changes like this are not common.

Once the foundry has acknowledged receipt of your design and confirmed design rule compliance, congratulations! You have taped out your design. Now the foundry can start processing the database.

A Note about Multi-Project Wafers

With new process technologies, mask costs tend to start out on the high side. In order to help reduce the cost burden to their customers and vendors, foundries are offering multi-project wafer (MPW) services. As the name implies, an MPW is a vehicle that merges many designs together on a single mask set, allowing multiple customers and/or vendors to validate functions and prototype designs. Since the aggregate cost of the mask set is spread over many users, each user pays only a fraction of the cost. Typically, MPWs are run on leading-edge technologies, 0.18um to 0.13um. In some cases, each technology node might offer both standard logic and mixed-mode or RF processes. Check with your foundry for any special MPW runs they might offer that are not on their current schedule.

MPW guidelines and schedules are available on the foundries' websites. Items you need to be aware of include:

First Come First Serve
You need to secure a reservation on the MPW well in advance. In some instances, the MPW run can be completely booked two to three months in advance.

Die Size Limitations
Typically, the foundries will accept one die size (e.g. 5mm x 5mm). Possibly, the foundry might allow you to buy four sites to accommodate a 10mm x 10mm die size. It is too challenging to dice many different chips with many different die sizes.

GSII Hand-off Deadlines
Since many designs must be merged together, the foundries need time to integrate the designs together on the same mask set.

Payment
If you sign up for an MPW run and do not use it, you will probably need to pay for it.

Foundry Deliverables
There is no yield guarantee. Expect about 50 to 100 diced die (not packaged units) meeting WAT testing. The foundry can drop-ship the die to your assembly or package vendor.

Cycle Time
Cycle time varies based on the technology node and the maturity of the process.

Design Rules
Make sure you are using the most recent design rules and SPICE models to avoid being dropped from the MPW run.

Again, the best advice is to work closely with the foundry to make sure you have the most updated information.

Phase Three: Mask Generation and Production Start

With tapeout approval, the foundry will get busy and add a seal ring, CD alignment marks, scribe lines, test keys, dummy fills, diffusion, metal, and a frame layout to your database.

This layout will then be sent to mask tooling. In most cases, the foundries will use subcontractors for mask tooling. (Even those foundries that have in-house capability will also have a subcontractor to ensure capacity.) Mask tooling adds data fracture and mask alignment marks and creates the MEBES mask data files. First masks typically take three to four days to complete, but expect longer lead times if your data preparation requires special handling such as OPC (optical proximity correction) and PSM (phase-shift mask).

In manufacturing, all PCM data are monitored by test keys inserted in the scribe line. Keep in mind that if your design requires extra test keys, the extra drop-in will impact cycle time and throughput. Foundries can add double or triple scribe lines if the customer requires special test structures that do not fit on a single scribe line.

Reviewing the Job Deck

Once the masks are completed, the mask house will check overlays of all layers for gross problems. At this point, the masks are done, and the customer has the option of viewing the job deck, or finished mask set, prior to making masks to catch errors. The "job deck view" refers to reviewing the job deck and checking to see that the mask set is free of errors.

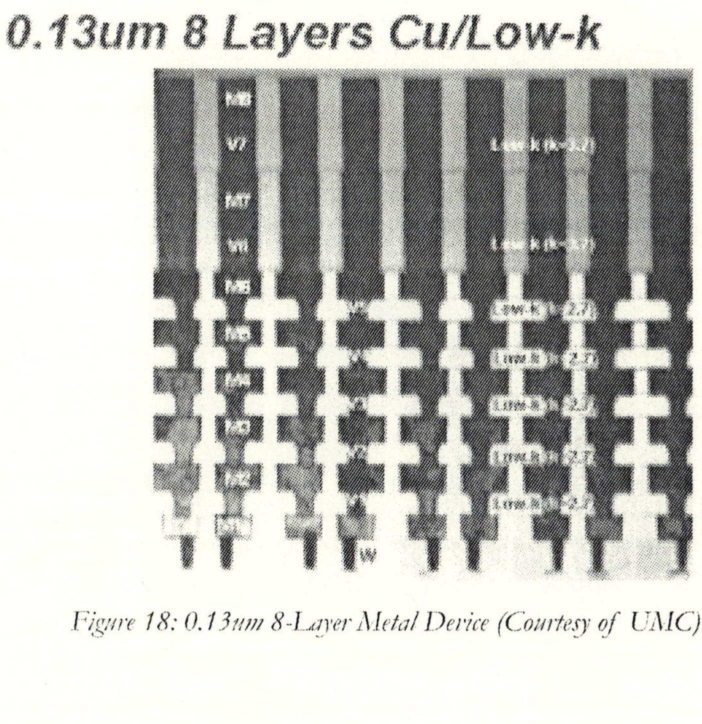

Figure 18: 0.13um 8-Layer Metal Device (Courtesy of UMC)

Reviewing the job deck is the last viable step at which the customer can validate that the GDSII data was converted properly to mask data. Unfortunately, not many people are well versed in how to review job decks. Key items to check include:

- The rotation, existence and polarity of data,
- The expected fractured data geometry based on CAD sizing and mask process bias, and
- The calculated geometry against a job deck.

The actual sizing of the job deck may differ from the data, due to biasing of the data. Work closely with the foundry to understand the correct measurements on the masks. Provided that they use the same mask size, you can check overlays, for example checking metal 1 and via or metal 1 and metal 2 jobdeck as a gross check.

The job deck is reviewed visually, usually on-line at a foundry office location. It is a good idea to have your lead design engineer with you when you go to review your job deck. Ask the foundry for a job deck checklist to help ensure that your design is ready for production.

Since the job deck is reviewed visually, you will not be able to spot small problems; you are only looking for gross errors at this stage during a job deck view.

While we have rarely seen this in practice, an idea to help you ensure that your job deck is in good shape is to enter a geometric or identifiable pattern in your design that appears only when your masks line up properly. Once again, a visual test pattern may not indicate detailed problems but it allows a quick check for gross errors during the job deck review. Once you are happy with the job deck files, the foundry will ask you to sign off on the design once again.

At this stage, the tapeout process is complete and your design is on its way to becoming a prototype. Now it's time to ensure that your prototype and production plans are executed.

Holding Wafers for Debugging

As your prototypes work their way through manufacturing, it is a good idea to "hold" a few wafers from the proto-lot "at metal," meaning that wafers are taken out of the production process right before metal layers are applied. This is done to speed up the prototype debugging process. Say that the prototypes in your first batch have some functional problem due to a design glitch. If you are lucky, it is a problem that requires a metal change only. By having wafers waiting at metal, you can speed up a new revision since you don't need to rebuild from scratch. Obviously, this technique will not help you save time with problems that require a full mask set change.

Keep in mind that the foundry will hold material only temporarily. Make certain you communicate to the foundry the expected time you will need for your debug. While there are no hard rules on holding material, if the foundry doesn't hear from you after four weeks, they may scrap your material. Stay in communication with the foundry to ensure this does not happen in the event you need more time.

Typical Foundry Forms

Information Release Authorization

Bank Name:__________________________________
Account #:___________________________________
Street:______________________________________
City: ____________________________ State:______
Phone: __________________________

Your signature below authorizes Foundry, Inc., to perform credit checks and to obtain credit history and performance from your company's bank.

Signature: ________________________ Date:________

FOR BANK USE: Foundry, Inc., would appreciate the timely response to the questions below relating to the credit history of Fabless Corporation whose authorization to release financial information is provided above.

Checking Account No.:_________________________
Date Account Opened:____________________
Current Balance:_______________________________
Average Balance:______________________________

NSF Activity
Date of Last NSF:_____________________________
NSF Amount:_________________________________

Remarks:_____________________________________

Form 6: Bank Information Release Authorization

Foundry Service Request Page 1/1

Customer Information

Company Name:______________________________

Address: ____________________________________

Telephone/Fax: ______________________________

Contact Name:_______________________________

Contact Ext: _________________________________

Contact Email: _______________________________

Product Information

Product Designation/Name:____________________

Product Application:___________________________

Expected Product Die Size:____________________

Expected Tapeout Date:_______________________

Technology Information:_______________________

Target Process Technology:____________________

Design Rule and Version:______________________

Expected Number of Masks:___________________

Wafer Handling Information:

Wafer Sort Required?__________

Assembly Required?___________

Turnkey Required?____________

If you answered "Yes" to any of the above, please contact the foundry for additional information.

Form 7: Foundry Service Request

IP Usage Form Page 1/1

Customer Name:____________________________
Customer Product/Project Designator:____________
Customer Purchase Order Number:______________

Are any second- or third-party libraries or IP used in this project (yes/no)? ________

If yes, please use the space below to list all second- and third-party libraries and IP blocks used in your design:

IP Usage

Vendor Name	Part Number	IP Description	Remark
1.			
2.			

Library/Memory Usage

Vendor Name	Library Type	Technology	Remark
1.			
2.			

Signature:____________________________

Date:____________________________

Form 8: IP Usage

Mask Tooling Request Form Page 1/3

1. Customer Name:______________________________
2. Product Name and Designation:______________

   ____

3. Main Customer Contact:_______________________
4. Tapeout Date: _______________________________
5. Purchase Order Number: ______________________
6. FTP Directory and Filename:__________________
7. Process Technology:__________________________
8. Poly Layers:_________________________________
9. Metal Layers: _______________________________
10. Die Size (No scribe)(um2): ___________________
11. Gate Length (um) ____________________________
12. Die Seal Ring Information:

    A. Customer-Added
    B. Use Standard Foundry Seal Ring

13. PCM Specification:

    A. Standard Foundry
    B. Customer Modifications (Please attach modifications)

Form 9: Mask Tooling Request

14. Test Key Instructions Page 2/3

    A. Use Foundry standard: No / Yes

    B. Additional Customer Test Key for Scribe Line

15. Wafer Handing Instructions:

    A. Select One:

        Super-Hot Lot

        Hot Lot

        Normal Lot

    B. Wafer Start Quantity: ____________________

    C. Create Probe Card Wafer:

        No / Yes (specify layer)_______________

    D. Corner Splits: No / Yes (If yes, answer below.)

        Foundry Standard

        Customer Modified

    E. Epi: No / Yes

    F. Backlapping: No / Yes (specify)_________

    G. Hold wafers (specify layer and number)

        ____

Form 9: Mask Tooling Request (cont.)

Page 3/3

16. Data Window (specify):____________________

17. Database Format: GDSII or MEBES

18. File Size: ______________________________

19. Special Notes or Instructions: _______________

____

16. Please match CAD files to the Mask Description Table below:

Mask Description	Mask Number	GDSII Layer	Remarks

Form 9: Mask Tooling Request (cont.)

Major Points of Chapter 6

- Set up an account and establish credit with the foundry.

- Run SPICE models, LVS, and DRCs prior to tapeout as well as on the database that was sent by FTP for tapeout, to minimize the chance of errors.

- Be prepared to send a PO prior to tapeout.

- Use the mask tooling form to confirm the final requirements for your design and to cross-correlate CAD files with foundry mask set nomenclature.

- Take full advantage of the job deck view to check your design for gross errors.

- Remember to request an extra wafer for probe development.

- Remember to request up to three wafers to be held at metal in order to speed up the silicon debug process.

CHAPTER 7

Qualification and Characterization

Once your prototype chips are on their way, major challenges still need to be overcome before the devices become production-worthy. The first step is to make certain your prototypes are functioning and free of bugs (this may take some revisions). The next step is to qualify and characterize your chip optimizing yield, quality, and reliability in preparation for your production ramp.

Prototype Phase

The general idea with a proto run is to get a certain amount of a product through the foundry as fast as possible so that system-level tests, failure analysis, and test vector optimization can begin. Generally, a foundry will start 12 wafers in the proto run, and it will usually run an extra wafer free of charge to use in probe card development. But do not get too hung up about getting your prototypes to pass your production test at either wafer or final package level.

Blind Builds

Remember that time is critical in the prototype phase. The objective is to get silicon through the foundry, packaging, and assembly so that chips can be tested in your system. It is usually acceptable to forego production tests and order a "blind build" for the prototype run. This is a common strategy. After all, what good are wafer-level tests if you are not even sure the prototype works in your system application? Furthermore, the test vectors themselves are not validated until they are tested and characterized on silicon, so it makes little sense to hold up a proto run in order to run your "untested" test vectors. At most, you may want to perform ATE tests on the final package level just as a quick "heartbeat" test to see if your prototypes have a pulse. This strategy allows you to quickly root out possible catastrophic problems such as electrical shorts in the circuit.

One consideration when doing a blind build is that you have the potential of losing valuable information if you run into problems. Visually looking at wafers and wafer-mapping sort yield can provide good clues when analyzing potential defects. At least keep track of wafer numbers and original die locations to help correlate with your WAT data. The inability to identify where a problem is might have more impact on time to market than the original time you thought you would save by doing the blind build.

Obviously, production test optimization can work in parallel in a proto run, even in a blind build. Bigger IDMs often minimize time to market by having production test personnel and system and design engineers working in parallel to bring up new silicon.

Remember to request some wafers to be held at contact. Work with the foundry to make certain they continue to hold these wafers while you work failure analysis of completed chips.

Software and Hardware Platforms

You will want to make certain that your reference design and any associated software builds are completed prior to the expected arrival of your prototypes. This point may seem obvious, but we find that many companies are not ready to accept completed chips. This is especially true today where massive SOCs require that many groups in a company work together to finalize the necessary software and hardware platforms prior to prototype arrival. (If your design requires the coordination of many groups of software and hardware developers, some basic project management rigor is probably going to be necessary for success.)

Analog Tests

If your device has analog or mixed signal blocks and you don't have a track record in analog design, or the design is radically new, you may want to consider running an analog test chip through the foundry in advance of your proto run using an MPW. Of course, this chip would have only the analog circuitry of your design.

If time allows, it may be good to run your analog tests prior to your general proto-lot in order to correct any analog problems that may crop up. You certainly do not want to derail your proto run with bad analog blocks. If, however, you simply do not have time to run your analog block(s) through first, then it still may be a good idea to use DFT to isolate the analog block(s) in order to determine their performance and serve as a better model from which to perform failure analysis.

Logistics

As your prototypes run through the factory, it is a good time to make sure the logistics of your process flow are in place. For example, if your package requires lead frames, make sure that you order these frames with enough lead time to meet the prototypes coming through assembly. If you are using a BGA package, keep in mind that PCB substrate may be on allocation. Make sure not only that the design is functionally validated, but also that material has been ordered in time to meet your chips coming from the foundry. Lack of planning for these components is problematic in a proto-run and potentially catastrophic in a production run.

As part of this effort, make certain that each vendor is prepared to drop ship products to the next vendor. This is fairly straightforward if the output of the foundry is going directly to an assembly house that performs all back-end services. However, it still makes sense to confirm that the foundry has the address and delivery instructions for your wafers. It is also a good time to review with the assembly house or your set of subcontractors the logistical steps to ensure your process goes smoothly.

Use of a semiconductor logistics provider is recommended to ensure parts are transported in a timely fashion with full visibility and communication to all concerned parties. A semiconductor logistics provider (forwarder) will work with your foundry and assembler to closely track and coordinate foundry release time parameters with next flight out, customs clearance, and assembler or test receipt. (See more on shipping and logistics in the next chapter.) [1]

Your vendors are not the only ones who need to be ready to receive parts. You also need to be ready to receive prototypes in your lab and begin the process of functional validation. This means, at a minimum, having the application test board ready to go, in time for prototype arrival. It also means having any firmware or API software ready to run on your chip.

Customer Support

Usually, you will have a customer who will want to start functional evaluation as well. The checklist below summarizes some of the major steps required for customer support at the prototype stage. Use this list only as an aid: satisfying it is necessary but not sufficient, as your own design may have extra steps not covered below.

Customer Support at Prototype Stage

- Applications Support
- Evaluation Boards
- Packaging of Boards
- Documentation
- Customer Deliverables

1 Contribution by Teresa Stewart, JSI Shipping, email interview, May 15, 2001.

Revisions

Functional validation is the primary focus in the prototype stage. It is hoped that any bugs found in silicon can be worked around in microcode or by doing a revision in metal. A revision of only your metal layers should not require a large lead time from the foundry since at least two or three wafers are waiting at metal in the process. However, there are circumstances where neither microcode nor a metal revision will solve a bug or functional problem. In these cases, you need to order a completely new wafer lot. Obviously, you want to avoid this because of the cost of generating a new wafer set as well as the cycle-time hit in building up new wafers. But sometimes it cannot be avoided.

Once you are satisfied that the functional behavior of the silicon is sufficient to meet your requirements, you are ready in earnest to qualify and characterize your device for production.

Qualification and Characterization

Testing, quality, and reliability are complex topics that are handled by specialists with years of technical training and experience. These specialists are in high demand. At the time of this writing, it is very difficult to find senior test engineers, as most are gainfully employed. This chapter will not replace the need to obtain or subcontract this expertise, but it will serve as a general guideline and highlight the major issues related to the topic of characterization, qualification, and reliability.

Manufacturing integrated semiconductor circuits reliably is an enormous technological challenge. This is no surprise, considering that devices today are built with gate oxide levels less than

20 angstroms, effective gate lengths less than 0.13 um, several layers of metallization, and millions of transistors. It is truly one of the giant leaps for mankind to get these complex devices to work at all.

In the COT model, the foundry is responsible for the quality and reliability of the manufacturing process. However, you are responsible for your product's quality and reliability. This requires correlating your product's yield and sometimes its performance to the process technology.

At a very high level, there are three issues that a product engineer deals with to achieve product validation for devices. These are quality and reliability assurance of the product, design robustness, and yield optimization. The end goal is to maximize the yield and reliability of your parts.

Quality and Reliability Assurance

It is important to recognize that the quality and reliability of a product is affected by its

1. Design,
2. Wafer fabrication,
3. Test methodology,
4. Packaging and assembly process, and
5. Final test and shipping.

Reliability is not hugely affected by step 5, and steps 1 and 3 are only of relative importance. Quality and reliability are primarily determined in steps 2 and 4, the wafer fabrication and the packaging and assembly process.

Foundries work hard to take a given process technology and make that process as stable as possible. When introducing a new process technology, a foundry will make continuous improvements by means of process characterization, as well as statistical control of quality and reliability.

The objective is to minimize the defect density and maximize the yields of the wafers going through the process, without giving up on quality. A foundry has an incentive to maximize its throughput of good wafers. They make money by selling good die to customers. Therefore, they will work hard to stabilize a given process technology in order to maximize the number of die they can sell.

Because of this stabilization process, design rules and SPICE models are often changing at the beginning of a new process technology. That is why it is so critical when you are working in the latest process to make sure your chip design and IP are keeping up with the foundry's tweaking and adjustments. Once the process is stabilized, and the defect densities and parametric variances have been adjusted to the limits the process can take, the design rules and SPICE models for the most part remain stable.

Reliability testing might include temperature cycling, humidity testing, ESD, and latch-up. These services can be provided by the foundry, by test vendors, or by outside reliability vendors. A design's robustness can be tested by the use of corner lots. Your design will also be susceptible to process variations. Performing a split lot will help determine optimum process parameters to maximize yields.

Mass Production Agreement

PCM Parameters

Once you reach a point where you feel you are ready for mass production, you need to agree with the foundry on the PCM parameters your production process should target. This agreement is normally done through a mass production agreement, an example of which is provided at the end of this chapter.

Before signing this agreement, make sure you spend some time reviewing the foundry's PCM specifications for your process. As a start-up company with a new product, it is unlikely that you will be demanding tighter PCM specifications from your foundry, unless your design has a justifiable need. PCM specifications will become much more important as your product goes into high volume and you find that certain parameters have direct impact on your yield. In the beginning, however, you simply want to ensure that your PCM specifications are not grossly out of line from your expectations.

At a minimum, you should review the major parameter specifications such as threshold voltage, saturation current, and resistor values. The parameters that the foundry presents to you will probably come in a table format, listing parameters, their respective sizes in width and length, a target specification, and an average deviance from the target specification. Measurement methodology may also be provided.

Once you review the PCM specifications, you will be ready to fill out the mass production agreement. The agreement's first order of business is to establish a set of PCM specifications. As mentioned above, a company will normally accept the standard PCM specifications provided by the foundry.

Yield Targets

Once you have agreed on a PCM specification, you need to agree with the foundry on a manufacturing yield target. Keep in mind that these yield targets can work in favor of you or the foundry. For example, if the yield falls below the acceptable threshold, the foundry will agree to pay for a new build, or at least provide a discount on your production lot. Section 7 of the form outlines in what instance the foundry will accept a "return," meaning no charge if yield fails to meet the target. However, if the foundry exceeds the minimum threshold, you can expect a premium charge on the wafer lot.

As a customer, you will also need to agree with the foundry that you will follow certain procedures and not hold the foundry liable for any problems with your product unless the problem stems from manufacturing. In addition, you will agree to maintain compliance with the foundry's design rules.

Once the agreement is signed, you are cleared for mass production. (If all goes as hoped, at this point your customers will be sending purchase orders for large quantities of your product and your production will start to ramp into high volumes.) Now issues such as yields and the bottom line become your major production concerns.

Correlating Defect Densities and Actual Yields

The product manager should be constantly correlating the defect densities of the process technology with the actual yield. Ideally, the yield of good die at sort should be equivalent to the theoretical yield possible at the current level of defect density. If your design is running into problems, you can ask to manipulate the process, for example by changing doping, changing mask skew, or buying a new mask set. If your design is susceptible to low L-effective, you could skew the mask down by a hair, though this sort of manual manipulation should be avoided if possible.

A yield which is less than expected for a given defect density, points to problems elsewhere in the process. A look at how "bad die" are distributed on the surface of the wafer can provide clues. If a problem is showing up consistently in the same location on a series of wafers, you might have a problem in your mask set. A random distribution, however, could imply a process problem such as high defect density.

It is not hard to correlate foundry performance with the yield and throughput of your products. Information on the wafer gate oxide, sheet rho, t-oxide, and effective channel lengths is usually available and can be used to correlate with your design. Most foundries provide websites that publish the status of the manufacturing process including PCM data, yield information, and other reliability statistics.

THE COT PERSPECTIVE

Analytical Laboratory Services

Sandra Delgato
Director of Engineering
Accurel Systems International Corporation

One of the most important alliances you will need to establish is the service of a comprehensive analysis laboratory. This should be done very early in your process development stage. Select one that can work with you at each step--from first silicon all the way to field failures. (Yes, even though you do your best, this can happen.)

What should you look for in a lab?

The lab should have a complete array of analysis tools and qualified personnel to perform each stage of the analysis. You want as much of the analysis as possible to be performed on-site at the lab of your choice.

The laboratory should offer the following services.

Secondary Ion Mass Spectrometry (SIMS)
SIMS can help qualify the ion in the foundry. It can measure small differences in dopant concentration (<1%) and distribution, resulting in fast and precise qualification of new implanters or processes, thereby shortening your overall time to product.

X-Ray Diffraction (XRD)

XRD can be used to check the orientation of the starting material to ensure it is correct. In dense line arrays, it can check the orientation of the grains in the metallization to see the strength of the fiber texture and ensure that all grains are oriented the same.

Atomic Force Microscopy (AFM)

AFM can measure the roughness of the surface of thin films (metals, dielectrics, etc.).

Field Emission Auger Electron Spectroscopy (FE-AES)

FE-AES is used for identifying submicron contaminants and thin film contaminants, for bond pad analysis (when experiencing bonding adhesion problems), and for confirmation of thin film composition.

X-Ray Photoelectron Spectroscopy (XPS/ESCA)

XPS/ESCA can also be used to identify materials, but due to the larger spot size (as compared with FE-AES), it is best for package materials (plastic or ceramic), plastic films, and some thin-film compositions.

Focused Ion Beam (FIB)

FIB is useful in the early stages of product development since it can isolate areas, cut connections, and make new connections, allowing you to fully test the chip before committing product. In the later stages of your product, it can be used for failure analysis by drilling down to lower layers and bringing up probe pads to easily test lower levels of circuitry and to isolate particular circuits for characterization.

Accurel Systems

Dual Beam FIB (DBFIB)

DBFIB is useful for failure analysis since it can precisely cross-section the smallest features while keeping the rest of the device intact. The Scanning Electron Microscope (SEM) in the dual beam allows nanometer imaging of the defects you are analyzing.

Transmission Electron Microscopy (TEM)

TEM is used to see the microstructure of different materials (crystalline or amorphous; single or polycrystalline). The specimen is made thin enough to allow electrons to pass through, yielding atomic-level resolution. The sample preparation can be done in several ways depending on the material and what you are trying to analyze within that sample. The excellent resolution (< 0.2nm) of TEM allows the analysis of very small features. TEM is also used to

- qualify different process conditions and make critical measurements of features,
- analyze the micro-uniformity and defects of the very thin gate oxides currently in use for advanced semiconductor processes,
- reveal failure mechanisms, and
- examine defect sites.

Electrical Characterization

Electrical characterization of your first silicon is key to understanding failure mechanisms, and it can lead to your long-term success. An e-beam prober is often used for the electrical characterization of your product. An e-beam prober is a specialized Scanning Electron Microscope (SEM) with a column that is optimized for the dynamic voltage contrast effect and a chamber arranged to facilitate biasing and exercising the sample. Beam blanking and filtering of the secondary electron signals are used to acquire voltage vs. time wave forms from the devices of integrated circuits, while they are being exercised. This probing is done without physically contacting the sample and with no significant loading being added to the signals. Typical applications would be following the data through a logic path or comparing the relative timing of different signals on a device for the purpose of debugging or characterizing the internal workings of a circuit.

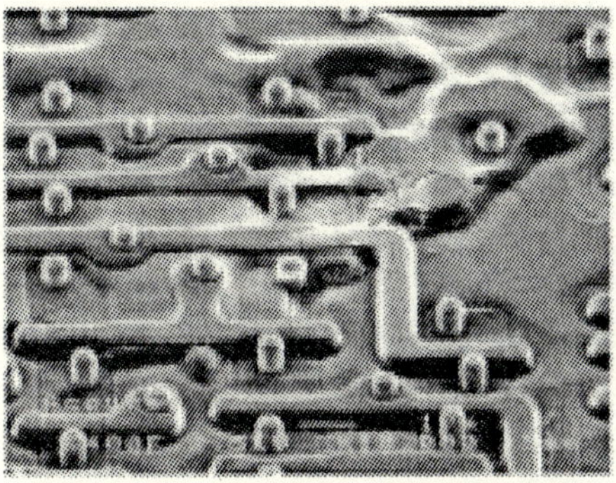

Figure 19: Tilted Close-up View of Diffusion Defect under Polysilicon (Courtesy of Accurel Corporation)

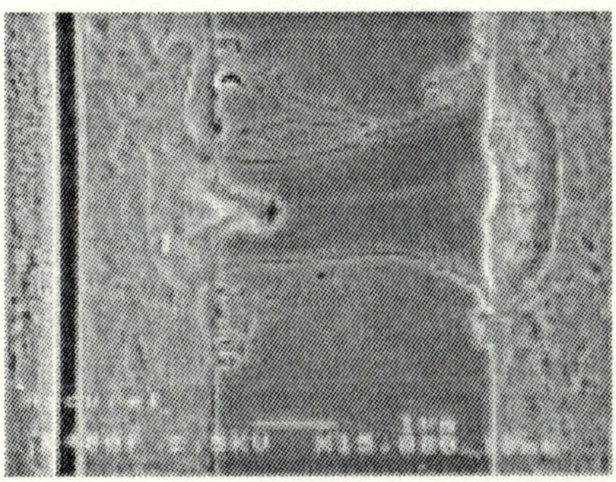

Figure 20: Close-up of ESD Damage from Contact-to-Contact across Resistor Structure (Courtesy of Accurel Corporation)

Non-Destructive Tests

Non-destructive tests such as Real-Time X-ray and Scanning Acoustic Microscopy (SAM) are valuable tools to see if the finished devices are constructed correctly (X-ray) and as a before and after check when running reliability tests (SAM). The laboratory can serve as your quality control check by performing tests such as bond pull tests, internal visual inspections, scanning electron microscopy inspection, and glassivation integrity testing. The lab should have the equipment and personnel to track down an exact failure location—Scanning Quantum Interference Device (SQUID) Microscope, Photon Emission, Liquid Crystal, Charge Induced Voltage Variation Analysis (CIVVA), etc. This simplifies the FA process, allowing the laboratory to deprocess a small area which can then be done quickly and with excellent results.

A good lab can help you qualify your product at every stage of product development. If you establish this relationship in the beginning, your laboratory will be familiar with you and your products at every stage of development and manufacturing, leading to reduced time to market.

The COT Perspective

THE COT PERSPECTIVE

Prototype to Production

Tom Napier
Test Engineering Services Business Manager
SABER

Frank Sauk
Diagnostic Services Business Manager
SABER

Introduction

The emergence and increasing complexity of system-on-a-chip (SOC) designs have created significant challenges for design validation. In the past, IC companies could decide to have an early tapeout followed by another design cycle in the case that the design did not function properly. But in today's highly competitive semiconductor industry, where time-to-market seems to be one of the most critical factors for success, there is no space for unnecessary tapeouts that require additional time and expenses. Thus, the steps and the design, test, and production engineers involved in bringing your prototype IC into the market are very important because they can make or break a new design.

The Prototype-to-Production Process

Once you have received the prototype IC, a smooth prototype-to-production flow moves from functional validation of the prototype IC, to characterization, and finally to ramping into production. If the prototype IC functions according to design during the functional validation of your design, you can then proceed with characterization that determines the operating range of your device as well as the amount of margin in the operating range to the device-published specifications. After that, development of a test program optimized for production and deployment of this test program to your test-house vendor of choice ensures yield and profitable binning.

If the prototype IC does not function properly, an additional design cycle will be required. (This occurs in almost 20 percent of the cases.) Instead of moving forward with characterization, your prototype IC will need to go through diagnostics in order to trace the root cause of the failures and repair the design. IC diagnostics let the design team validate the root cause of the failure before you commit to expensive mask changes and a second tapeout. After the design revision, you can move forward with the mask changes, silicon fabrication, functional validation, characterization, and finally ramping to production.

The COT Perspective

This section covers the prototype-to-production process flow that is depicted in the figure below.

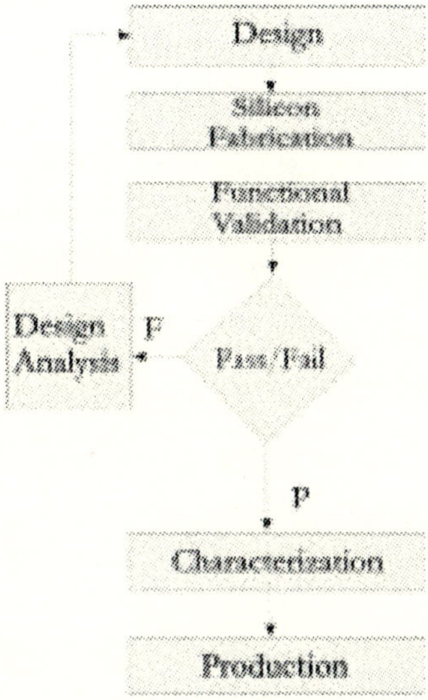

Figure 21: Prototype to Production Flow

Functional Validation

How quickly the design is validated and any identified problems resolved will affect how soon the prototype IC can ramp to production and be brought to market. Therefore, this process directly affects the time-to-market of your IC. Functional validation consists of several steps:

1. Hardware Verification

Once the test hardware is built and the loadboard test is written, the test hardware should be verified. You need to make sure that the loadboard test passes and that the test system passes the hardware diagnostic before you attempt to validate your prototype IC.

2. Continuity Test

The continuity test is sometimes called a contact test or opens and shorts test because it verifies that contact is being made to all signal pins and that no signal is shorted to another signal or to power or ground. The continuity test can be used to check that the paths between the tester interface and the pins are not open and that the paths to the pins are not shorted together, shorted to ground, or shorted to a power supply.

3. Gross Current Test

The gross current test is much the same as the continuity test for the power supply pins. After you initially power up the prototype IC, sometimes with a low voltage, this test checks that it is not drawing excessive current that may damage the test hardware if testing were to continue.

4. Functional Test

This test is the heart of the functional validation process. It verifies that the prototype under test correctly performs its intended logical functions. To accomplish this, test vectors or truth tables must be created in order to detect faults. This is not a full characterization of the device; that step follows next as long as the basic functionality of the design is validated here.

5. Structural Tests

Structural tests can decrease test development time as well as test time. Structural tests depend on designing special test circuits into the ICs. Design for test (DFT) is a structured design methodology that greatly reduces the complexity of the functional test requirement. Scanning chains is a popular DFT approach that provides a way to test parts of the IC in shorter time. Logic built-in self test (BIST) is another approach that is particularly effective for embedded memory that requires only deterministic testing.

The COT Perspective

Functional validation of your prototype IC validates that your design performs according to the design models. It is a step that every IC should go through before it is moved to production. Thus, an efficient test strategy with a validated test methodology, and a complete set of test vectors handed to your test outsourcing partner, directly affect the functional validation process and the time-to-market.

Diagnostics

Fabless companies that want to beat the time-to-market challenges for their designs face several decisions early in the silicon development cycle. Decisions on EDA tools, foundry services, and manufacturing processes, as well as on test and assembly houses, are usually taken early, up-front in the cycle so that design teams can work efficiently towards tapeout and prototype IC fabrication. However, once first silicon is received, it is very common for designers to assume that the prototype IC will perform according to specifications when it goes under test.

Therefore, in a lot of cases, companies do not have a debug strategy in place when the prototype IC is not functionally validated. A widespread approach is to go back to the EDA tools and use extensive simulation to trace the fault on the design. Split lots in the foundry that simulate possible faulty behaviors follow this approach. The test equipment can also be used to characterize the device and produce shmoo plots that could indicate the error, or to take measurements on the prototype IC. Analyzing and comparing these measurements with the simulation corner predictions could point out the fault in the design. Depending on the complexity of the device, this procedure can become very hard, especially if the fault is not direct, because it is not correlated to an identifiable fault, which is normally the case. As design complexity increases, effective simulation requires a significant time period and mask changes become even more expensive.

The trade-off between accuracy and speed is always a big concern when you use computer simulation to debug your prototype IC. Diagnostics technology offers another alternative for your debug strategy; it allows you to trace the root cause of the failures with waveform measurements performed on appropriate probe points on the prototype IC itself.

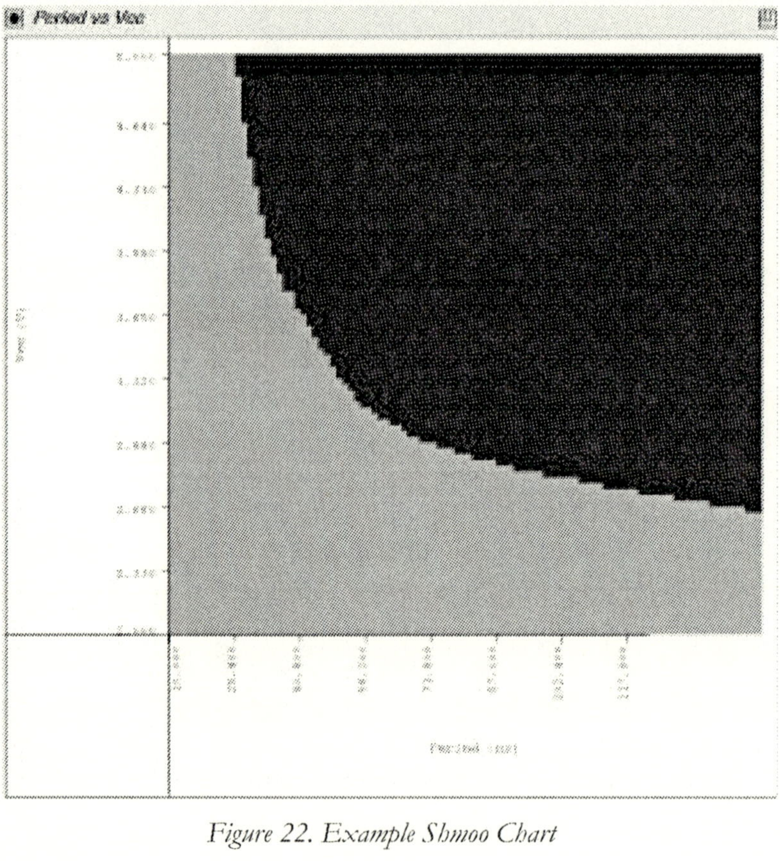

Figure 22. Example Shmoo Chart

The COT Perspective

Several diagnostic technologies were developed over the past few decades, evolving from mechanical probing to electrical probing and finally to optical probing. Some of the most common mechanical and electrical probing technologies are the following:

- Mechanical
- E-Beam
- Combination Methods
- Atomic Force Microscope (AFM)
- Focused Ion Beam (FIB)

The latest evolution of probing technology is optical probing:

- Laser Voltage Probe
- PICA
- Emission Microscope
- Flip-Chip Focused Ion Beam (FIB)

Focused Ion Beam (FIB) technology is an enabler for diagnostics, rather than a technique which provides useful results by itself. It is used to drill down to other layers, so that E-beam, for instance, can continue tracing the fault and measure waveforms on probe points on lower layers. It is also used to repair silicon once the root cause has been identified to allow functional validation of the prototype IC before a new design spin. The FIB may also be used to generate a number of samples to customers before a new design spin takes place.

The optimal debug strategy depends significantly on the type of failure observed as well as on the process technology used in the manufacturing process. For leakage failures, emission microscopes and liquid crystal technology are the most applicable.

For marginal performance, with race conditions, process corners, or temperature dependency, various technologies may apply. Mechanical probing and atomic force microscopes are recommended for DC and very small measurements in the millivolts. Mechanical probing and atomic force microscopes work well for intermittent failures when looping becomes difficult. E-beam, laser voltage probing, and PICA are recommended for timing analysis with simulation feedback. For packages with standard interconnects down to .13u, E-beam is the most efficient technology. Laser voltage probe can go further down to 0.10um for both standard interconnect and flip-chip packages. Finally, PICA is used for silicon-on-insulator (SOI) process technology and is applicable primarily for failure analysis of devices.

Simulation has computational limitations, especially when failures are non-systematic and cannot be modeled. Simulation results can be only as good as the models provided. In contrast, the advantage of diagnostic technology is that the waveform measurements are taken from the prototype IC itself and not from a computerized model. Thus, you should view diagnostics as an effective insurance policy that can control the risk of delay to market.

Characterization

Due to today's elevated frequencies and high pin counts, signal integrity has become especially critical. Reliability and repeatability are essential for comprehensive characterization of specifications. This test is a method of distinguishing the particular traits of a prototype IC for various operating parameters. The data collected are presented in a meaningful way from which certain conclusions can be derived.

Characterization consists of eight steps:

- Loadboard validation
- Loadboard open circuit test without the prototype IC
- Opens/shorts test on prototype IC
- Functional test: Test vector patterns designed specifically to characterize a single device function are needed in order to collect meaningful data from a test. The function of the test vectors needs to be clearly understood before being used for characterization, especially for AC parameters.
- DC parametric tests: Characterization of DC parameters can extend to tests such as leakage current on certain pins. If these attempts are successful, it will be possible to identify and localize internal physical problems that are causing the device to fail prematurely. This kind of information is not available through functional testing. Parameters usually characterized on patterns and frequencies chosen are

    - VDD
    - VIL/VIH
    - VOL/VOH

- AC parametric tests: AC parameter measurements can be taken through a binary search, also known as successive approximation, or through a linear search. The difference between binary and linear searches is that during the linear search you move the voltage, current, or timing unit in a single continuous direction. AC parameters usually measured are

  - Setup Time
  - Hold Time
  - Propagation Delay
  - Maximum Frequency

- Temperature parameters tests: The prototype IC is stress-tested at typical temperatures of 0 degrees Centigrade to 100 degrees Centigrade. Testing the IC in these extreme temperatures validates the quality of the manufacturing process.

- Operating current: The operating current of the device needs to be characterized over its entire voltage, frequency, and temperature operating range. The characterization of the operating current should be done on all bias to the device and in all operating modes (i.e. standby, sleep, and functional).

Exhaustive tests are performed to determine the IC operating area independent of the actual specifications. Various parameters are tested and the interaction between parameters is evaluated. An ATE is required in this application to perform the millions of tests necessary to characterize the IC accurately and completely.

The data collected during characterization have many uses. The manufacturer uses the collected data in order to establish or verify published specifications. The data are also used to assist in some applications of the IC and in research, product marketing, quality control, and reliability studies. For the fabless company, the characterization data are an invaluable tool in helping the design engineer develop the final product design. And characterization provides information for identifying potential

The COT Perspective

sources of additional profits, from ICs that are classified as better than the specification. This information is critical to performing yield management, helping you quickly identify problems and predict performance of your IC when you ramp to production.

Production Test

Once your prototype IC is functionally validated and characterized, it is ready to ramp to production. Programs written for production test are intended to test newly fabricated devices to make sure they meet the specifications determined by the manufacturer. An ideal test program rejects all devices containing defects and does not reject good devices. In general, the goal of this type of program is to comprehensively test the IC in as short a time as possible for the least possible test cost per device. Test time needs to be kept as short as possible as it directly affects the cost for each device tested. For instance, to test continuity in a production mode, a functional continuity test is often used instead of a DC test for better throughput.

The production test program should be validated and released for deployment to your production site. Deployment consists of installing the production test program on your production equipment, performing correlation to anticipated results in accordance with your specifications, and following the release process of the test house.

Tips for an Efficient Prototype-to-Production Process

An efficient test methodology helps reduce the validation and characterization process. Test program development usually takes around six weeks. The ideal time to engage an outsourcing partner for back-end test engineering is two months before tapeout, to allow for validation of the test strategy as well as for development of the test programs and hardware required. The test platform and the interface solution should be decided after your device pin list is frozen. (For tips on ATE selection refer to Chapter 5: Packaging, Assembly, and Testing.) If you have already decided on the package of the device, it is very useful to ask for a dummy package to evaluate the socket, since there is always risk involved in designing the socket for the device.

The value of the test program will depend on the purpose for which it was written. In general, test programs are written for one of three purposes: device validation/debug, characterization, or production. A single test program can be written that will fulfill all of these purposes; however, one should be careful to partition the program to accommodate each purpose without compromising the others.

Test vectors and the signal timing associated with them are very important. The test vector set used for functional validation should be complete in order to meet your quality goals for returned devices from your customers. Missing vectors lead to unidentified problems, or problems that cannot be resolved because your test outsourcing partner does not have a deep understanding of the device's functionality and only depends on the test vector set while testing the device. Test vectors can be generated by simulation and usually have to be converted to a format compatible with the tester. A conversion tool is required to perform this function. The source of the vectors needs to be considered very carefully for both functional and structural (scan) tests.

The COT Perspective

Performing device characterization requires careful planning. The device specification must be reviewed to develop the list of parameters to be characterized. Problems are usually not present in DC parameters or input and output threshold. However, AC parameters could be critical to the application for which the device will be used, and they are not tested under a digital test system.

An experienced test engineer is required in gathering the desired data for device characterization because of the detail that is required. Consulting with the design engineer and communicating in the right language is very important in order to resolve any questions regarding timing relationships or device functions.

You can reduce the diagnostics process significantly with setup preparation before or during the tapeout. A typical diagnostics project consists of the design and manufacturing of a board, vector generation to stimulate the failure mode, CAD file processing to enable navigation on the IC using the layout information, and sample preparation. The design and manufacturing of a DUT board has the longest lead-time (three to six weeks), whereas the actual waveform acquisition could take only hours or a few days. You could prepare ahead of time in order to be ready to go through diagnostics immediately after the functional validation of your prototype IC fails.

Various methods of design rules can help you with the debug process, such as good alignment points, spare structures instead of just cells, accessibility, and changes. Design rules so that FIB can be easily applied to prototype ICs can reduce the design cycle time, increase the efficiency of the diagnostics cycle, and decrease your mask revisions.

Minimize Prototype-to-Production Process Time with Remote Collaboration

Once you have selected your foundry and your back-end vendors, you will end up with multiple suppliers located around the globe and a challenging supply chain to manage. At the same time, most small companies have resources focused primarily on the design and marketing of their devices. Even larger fabless companies that might have a separate group focused on test engineering often face limited resources for silicon validation, debugging, and characterization. In order to be profitable and to remain competitive, your global business environment requires widespread global partners to work closely together from design to manufacturing to accelerate time to decision for every step of the process so that you move your prototype to the market.

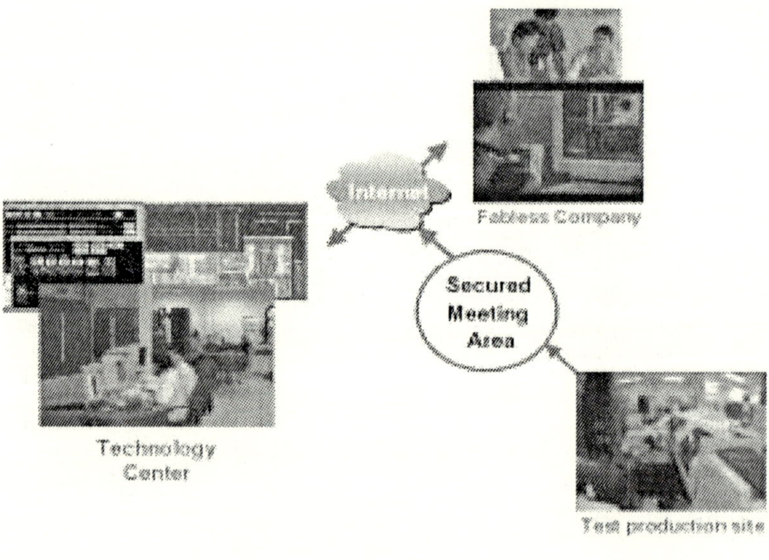

Figure 23: Remote Collaboration (Courtesy of SABER)

Real-time interaction with partners worldwide allows each company to focus on its core strengths along the development-to-production cycle. As a result, the time to design, validate, characterize, debug, repair, test, and deliver new products is the shortest possible and it comes at the lowest cost.

In order to achieve real-time interaction and collaboration with your outsourcing partner, you need a secure high-speed networking infrastructure and a collaborative environment with application sharing. That way, your design engineer can interact with your partner's test or diagnostics expert anywhere in the world, or access the equipment from your site, your office, and your desktop without having to travel. This means that once you have received your prototype IC, your partner could have already prepared the test environment to enable immediate validation. An operator will set up the environment for your usage, and with audio and visual assistance you can collaborate remotely with experts and interact in real time with them to get your problems resolved.

If you outsource the functional validation and characterization of your prototype IC, collaborating with your partner enables you to share validation or characterization results, oscilloscope capture, and timing windows. The end result is not only to speed up the process of validating and characterizing your prototype IC but also to debug it, in the case that a problem has been identified.

If your prototype functional validation fails and you decide to outsource the diagnostics process, your design engineer may interact in real time and point to the critical node to probe, either on the schematic or on the layout, to help identify the root cause of a device failure. Once waveforms have been acquired, your partner's expert can share them with your design engineer, who can then direct him how to move forward to the next probing session. If any particular knowledge on a specific design block is required, you have easier access to your key designers from your facility, which is extremely critical for a debug session where deep knowledge about the device behavior is required to identify the root cause of a failure. Real-time interaction helps you avoid probing unnecessary points and to isolate the fault faster. Therefore, a repair session can start sooner and you can deliver the first working sample to your customer sooner.

Through a secure network connection with foundries and test houses overseas, you will be able to follow up on your test solution deployment. In this context, when your partners deploy test solutions into production environment, you could manage remote correlation in cooperation with the team that had developed the test program. Remote collaboration allows you and your partners to deploy your test program into production, but also to manage new test program versions, investigate the origin of any yield losses, and improve the test cell. This capability can be used for production test at a test house or for wafer sort at a foundry or a test house.

Remote collaboration can reduce your prototype-to-production process cycle and therefore the time-to-market for your device by:

- Reducing travel time during ramping into production, which means that your design engineer's time is freed up to focus on new designs,

- Accelerating ramp-up time and test solution transfer, which translates into reducing time to money for you, your back-end vendors, and your customers, and
- Improving yield issues resolution, test cell performance, and test time per device.

MASS PRODUCTION AGREEMENT

1. Customer Name:______________________________

2. Customer Product Name: _______________________

3. Process Technology: __________________________

4. Mask version list in production:__________________

5. PCM SPEC: Standard

   Modified, NEW PCM Spec Attached

6. Average yield: __________________________________

7. Foundry accepts low yield return criteria:

   Percent nominal yield:______________________

   Percent standard allowance yield:_____________

   Standard compensation criteria by lot:________

   Standard return criteria by lot:________________

   Standard return criteria by wafer:_____________

Other Terms:

Return criteria effective after 1000 wafers from pre-production start.

No changes to test program without prior consent.

Form 10: Mass Production Agreement

Compensation amount is

(standard – actual) * quantity * price / standard

where standard is the standard allowance yield and actual is the actual lot average yield.

8. Foundry agrees that:

A. All procedures follow QA procedure.

B. All processes will be in compliance with PCM spec. (Attached is a copy of PCM.) If the customer changes the PCM specification, the foundry reserves the right to redefine the normal-yield and low-yield return criteria.

9. Customer agrees that:

A. Foundry has no liability for product reliability unless incorrectly processed.

B. If customer's layout violates foundry's design rules (DRCs), foundry has the right to reject the low-yield claim.

C. If yield is below agreed percentages, customer will follow established procedures to request compensation.

D. Customer will provide low-yield compensation requests within two months of wafer shipment.

Form 10: Mass Production Agreement (cont.)

Major Points of Chapter 7

- The general idea with a proto run is to get a certain amount of a product through the foundry as fast as possible so that system-level tests, failure analysis, and test vector optimization work can begin.

- Make sure that all the necessary material has been ordered to support the manufacturing of your product through assembly and test, such as lead-frames or PCB substrates. Lack of planning for these components is problematic in a proto-run and potentially catastrophic in a production run.

- As part of this effort, make sure that each vendor is prepared to drop ship products to the next vendor.

- You may consider outsourcing the planning and quality to a semiconductor supply chain management service.

- At a very high level, there are three issues that a product engineer is dealing with to achieve production-level products. These are quality and reliability assurance of the product, design robustness, and yield optimization. The end goal is to maximize the yield of reliable good-quality parts.

- Once you reach a point where you feel you are ready for mass production, you need to agree with the foundry on the WAT parameters your production process should target.

- The product manager should be constantly correlating the defect densities of the process technology and the actual yield.

CHAPTER 8

Planning and Logistics

Production Planning

The production planning that is necessary for wafer fabrication is just as sophisticated as any assembly process from automobiles to electronic equipment manufacturing. Volumes have been written about the theory and practice of orchestrating the inputs and outputs of complex processes. The management of production planning is a science unto itself.

Fortunately, your foundry and subcontractors take responsibility for much of the detailed production planning that goes into actually building an IC. However, you are still responsible for ensuring that you order enough devices to meet your customer needs when they need them. At first this may sound simple enough, but it can be challenging, especially as certain throughput parameters, such as yield, fluctuate with time. In addition you may need to orchestrate cooperation and lead times among several manufacturing subcontractors.

People going through the manufacturing experience for the first time are often ill prepared to deal with production planning. Many of those who are starting up companies come from bigger IDMs where the planning function is taken care of by staff who are well versed in the art of planning capacity for expected demand.

Software is available to help manage production planning. And for more established companies with multiple products and customers, a software management tool is indispensable. However, smaller companies may get away with foundry-provided e-commerce links such as wafers in progress (WIP) to do their planning.

You can also subcontract planning activity to a company that provides planning and quality services. Such companies have specialists on hand to manage these activities, along with software connected to all your foundry, assembly, test, and shipping vendors.[1]

The trick to production planning is to understand three basic concepts: your own company's demand forecast for your chip, the current yield, and the lead time through the foundry.

Forecasting Demand

Forecasting demand is not always easy. However, as a start-up you are wise to build only on purchase orders received. Purchase orders that your company receives will normally list the number of devices requested and the customer request dates for delivery. (Ideally, of course, you will have many purchase

1 Contribution by Teresa Stewart, JSI Shipping, email interview, May 15, 2001.

orders coming in requesting a great deal of product at a high price.) In this situation, you simply order the appropriate quantity from your foundry to meet your total purchase order requests minus your yield hit.

For example, say you have orders for 50K units of a product to be delivered in four months to your customer. In this simple example, assume that the total yield through the entire manufacturing process is 50 percent.[2]

Current Yield

If you know the die-per-wafer number provided by the foundry and the total yield hit, you have the necessary information to plan your production of 50K chips. In this example, each wafer carries 200 die. Given that your yield is 50 percent, you will need to order 500 wafers in order to expect a total of 50,000 good die to be available to your customer. (The math is 500 wafers x 200 die per wafer = 100,000 die; 100,000 die x .50 = 50K chips.)

From this example, one can see the importance of nailing down a good prediction of yield through the process. As discussed in the previous chapter, yield is not a constant—especially for new process technologies, which generally have higher defect densities. As a process matures, you can expect defect densities to go down and yields to go up. Defect densities are also a function of die size.

2 Detailed yield calculation examples are available in chapter 9.

Yield is also a function of process variations. Corner splits, explained above, would optimize your good-chip-per-wafer throughput.

Fortunately for the armchair planner, the foundry's e-commerce tools will make it possible to track defect density information. Together with knowledge of your die size, this should be enough information to predict yield and to keep up with yield changes throughout the life-cycle of your product.

Bose-Einstein Model

Most companies rely on the Bose-Einstein model to estimate yield for a given defect density and area. However, the Murphy model is also an option. Ask your foundry to specify which model and association parametric values they use.

The Bose-Einstein model (BE) is typically used by the foundries to calculate logic yield. The definition of BE yield is

BE yield = $1/(1 + AD)^N$, where
A = area of the die (inch^2),
D = defect density (defects/inch^2), and
N = number of critical layers in the foundry process.

D and N values should be provided by your foundry to help you calculate your device yield. It is important to understand that when calculating your yield, SRAM defect densities should be treated differently if the bit-cell of the SRAM block violates logic design rules. Work with your SRAM provider and foundry to help determine defect density for these blocks.

The table below gives you an idea of gross die per wafer (GDPW) and yield estimates for various die sizes using the

Bose-Einstein model. The example assumes 0.25um-process five-layer metal.

		Yields at Defect Densities 0.5 to 0.15				
Area (mm2)	GDPW	0.5	0.4	0.3	0.20	0.15
16.00	1814	83.4	86.5	89.6	93.0	94.7
20.25	1419	79.5	83.2	87.1	91.2	93.3
27.56	1026	73.2	77.9	82.9	88.2	91.0
33.06	847	68.9	74.1	79.8	86.0	89.3
39.06	709	64.4	70.3	76.7	83.7	87.5
45.56	602	59.9	66.3	73.4	81.3	85.6
56.25	480	53.3	60.3	68.3	77.5	82.6
81.00	323	40.7	48.5	58.0	69.4	75.9
100.00	256	33.3	41.2	51.2	63.7	71.2
126.56	197	25.2	32.8	43.0	56.7	65.2
132.25	187	23.7	31.3	41.5	55.3	64.0
156.25	155	18.5	25.6	35.5	49.8	59.1
175.56	135	15.2	21.8	31.4	45.8	55.4
196.00	119	12.4	18.4	27.6	41.9	51.8
203.06	115	11.6	17.4	26.4	40.6	50.6
240.25	94	8.1	12.9	21.0	34.7	44.9
289.00	75	5.1	8.8	15.6	28.2	38.3
306.25	70	4.3	7.7	14.0	26.2	36.3
400.00	51	1.9	3.8	8.1	17.8	26.9

Table 8: Bose-Einstein Model Example
8" GDPW and Yield Percent for (Do,n=14.75) for 0.25um 1P5M

When introducing a new process technology, foundries will provide estimates of defect densities over time. An example might look like the following, where Q1Y1 is an early production release.

Estimated Defect Densities over Time					
Quarter	Q1Y1	Q2Y1	3QY1	4QY1	H1Y2
Density	1.00	0.60	0.50	0.35	0.25

As you can see, the defect density improves over time as the process matures. A mature process like 0.18um technology would have a defect density around 0.2 defects/in^2. These estimates can be used to generate your pricing strategy and revenue forecasting numbers.

Foundry Issues

As a planner, do not forget to make sure that the foundry lives up to its commitment to provide the best production throughput possible for a given process technology. In the mass production agreement you sign with the foundry, you and the foundry not only agree to set the PCM data and certain key performance parameters such as L-effective and break-down voltage, but also agree to clauses that govern foundry compensation if these targets are not met. (As a planner, remember that the mass production agreement should also spell out major process controls as well as minimum hold and scrap requirements.)

A word of caution: if the foundry scraps material for any reason, never assume that they will automatically restart wafers for you. Make sure you place a new order in the system and track it to guarantee that your material is going through the system.

Along these lines, always keep tabs on how your wafers are moving through the foundry. Often we run across situations where a company submits its GDSII database and basically assumes that the wafers will be available at the delivery dates stated by the foundry. While the foundry has the best intentions to deliver your product on time, keep in mind that there are hundreds of customers' wafer boats floating through the factory at any given time. It is possible for wafers to get side-tracked for some reason, so make sure you are tracking the progress of your wafers in the WIP (wafers in production) report. If your material does not move over a reasonable amount of time from one stage to another, contact your foundry to make sure there are no problems.

A Planning Database File

Planning is more than just predicting yield and making sure that your product is moving through the foundry. Many dates and benchmarks need to be correlated and tracked in order to ensure a smooth flow of the process. This is where a database file can be of tremendous help.

In setting up your planning in a database file, you should track at a minimum the customer request date for the product, the projected actual date of delivery based on your calculated cycle-time and the foundry's commit date,[3] and the amount of product requested. You can use a database script file to download and track all kinds of yield processes related information.

3 These days foundries are using a new concept in supply chain management known as "available to promise." This system provides a way to place an order to the foundry and immediately receive capacity and delivery schedules.

WIP Information

WIP information for each stage of the fabrication process should also be tracked in your database. For example, there should be WIP dates and estimations for sort, assembly, and test. Like foundries, assembly houses provide their customers with access to e-commerce tools. Assuring that progress is being made in each category of a process is just as necessary in these stages as it is in the foundry stages.

Transit Information

Make certain to track transit times from the foundry to the assembly house, from the assembly house to your inventory, and from inventory to your end customer. Transit times need to reflect your particular process. The transit information should also include any waybill numbers and tracking numbers from the shipping company. You will need these numbers if your product is not showing up at a particular stage.

Shipping Companies

There are advantages to working with a company like JSI that employs semiconductor specialists. These specialists will not only track shipping data but also advise you of precise transit options as well as standard and exception factory release times to further enhance your cycle-time reductions. For example, if you are moving your product from Taiwan to the Silicon Valley, JSI offers an express service of 29 hours or less. They arrange early factory release from your foundry or assembler to permit same-day shipping out of Taiwan and same-day receipt by you or your test house in the USA, seven days a week. JSI's complete transit studies show precise factory release to factory receipt (with transit time inclusive of factory release, customs clearance, and factory receipt), on a standard and express basis for all global destinations, 24/7/365.[4]

Beware of shipping companies that quote same-day service and fail to back it up with details. It is important to understand what your shipping company considers the start time for calculating "overnight service." For example, one company may start the clock at factory release and end it at factory receipt, while another company may not start the clock until material is picked up or reaches some distribution point.

Another point to consider is the number of days service is available. Is the service offered daily, weekends, after hours, and on holidays? You don't want to pay a premium for a super-hot lot only to lose the time savings in extended shipping time. Ensure that your vendor can provide detailed transit times. Also, you should ensure that your shipping vendor is familiar with all your other vendors' receiving and shipping hours and standard/exception factory release times. Finally, you should ensure that your vendor can provide expedite services through customs.[5]

Below are a few additional guidelines on setting up an account with a shipping company.

- Open an account well in advance if possible. It reduces unneeded stress and prevents unnecessary delays. The following forms will need to be signed off.

4 Ibid., Stewart, JSI Shipping.
5 Ibid.

- Power of Attorney: Permits the customs broker to act on your behalf to clear your shipments through customs. (This requires a corporate officer's signature. Not all logistics providers offer customs clearance services.)
- Routing Order/Instruction: Informs your foundry and assembler vendors of your logistics provider.
- Authorization Letter: Notifies all airlines, forwarders, and brokers of your authorized customs broker.
- Credit Form: Your logistics provider should provide 30 days credit.

• Set up a continuous bond or single-entry bond. Your shipping company will arrange this on your behalf with U.S. Customs. Note: U.S. Customs requires 10 business days to activate the continuous bond. The advantage of a continuous bond over a single-entry bond is that it permits much faster U.S. Customs clearance and is more economical if you ship more than ten times a year into the United States.

• Review whether your corporate blanket policy covers your global shipments. If not, add an umbrella or rider to the existing policy. Another option is to purchase transportation insurance from your semiconductor logistics provider.

• Use a one-stop shop. Confirm that your logistics provider offers shipping, customs clearance, and pickup and delivery services. Most services provide only a portion of these functions during business hours. A semiconductor logistics specialist provides not only all these services, but also distribution and supply chain management services, 24/7/365.

- Request transit schedules. A good logistics provider will show transit from factory release to factory receipt with express options.

- Obtain export/import compliance services. Avoid high lawyer fees by working with a semiconductor logistics specialist that provides these consulting services at no additional cost to you.[6]

Packaging

You should also be aware that there are many ways to package your final chips for shipment to you or your customer. For your product chips that will likely sit in inventory, you will probably want to specify a bake and seal process with your package vendor.

You will also need to decide whether to pack your chips through tubes or by a tape-and-reel method to accommodate high production requirements. Check with your end-customer if you are not sure.

Whether tape-and-reel or tubes, the chips must be put in boxes for shipment. You must decide whether you need special barrier bags, desiccants, or humidity card requirements. You will also need to specify marking instructions for the outer box, including what you want on the packing labels. In addition, some companies require shipping and certificates of compliance to be attached to the outside as well as to inside containers for export and/or import.

6 Ibid.

The assembly house will also gladly mark your chips any way you like; however, you need to specify and request marking from the assembly house for this service.

Inventory Management

In an ideal world, you would never have inventory. The exact number of chips that you produced with successful yield and test passing would always equal the number of chips that your customers wanted.

But the world is far from ideal. Yields can vary and purchase orders can be canceled. Therefore, you may end up with more chips then you can send on to your customers at a given time. Customers may cancel orders, yields may increase due to process or new PCM agreements, and other factors may lead to situations where you have more devices than you can sell.

Subcontracting Inventory Storage

Inventory can be set up in your own facilities. However, for many smaller companies this is not feasible. Depending on the amount of products you need to store, inventory may take up a lot of room. Moreover, if you have inventory in your own space, you will also need to accommodate the coming and going of delivery trucks, which usually means a shipping and receiving department next to a truck bay. A company with a 500-square foot office in a high-rise building may not have the luxury of accommodating such activity.

Most likely, you will want to negotiate with a subcontractor to provide inventory for your finished products. Assembly and test houses, foundries, and shipping companies often provide these services for a fee. Obviously, the greater the inventory, the more it costs to store your products. Therefore, the goal is to minimize inventory while maintaining a reasonable buffer of products to meet unexpected demand upsides or yield losses. There are no hard and fast rules about the size of this buffer. Certain situations may dictate carrying higher or lower levels of inventory due to the nature of a particular business. In addition, inventory should be temporary, as many OEMs do not accept parts that have sat in inventory for over a year.

Storing WIP

Keep in mind that inventory does not have to mean storing packaged die. Inventory can also refer to storing products in wafer form. It is generally much cheaper to store WIP, especially if you need to store a large number of products at a given time.[7] Inventory costs increase as a product takes on size. Products in wafer form also have less value than they do in packaged and tested form, in addition to taking up less space.

You can negotiate with foundries and assembly and test houses as well as shipping companies to hold your WIP. WIP requires special handling, however, as well as clean-room storage facilities. Once again, the more inventory you have, the more it costs. You also have to ensure that wafers follow a "first in first out" flow through inventory in order to track these lots through the process. You should assume a shelf life of about one year for products in WIP inventory.

7 Storing inventory of material at the wafer level can save up to 60 percent of costs over inventory of final product.

Handling Perishable Inventory

The time element associated with holding semiconductor devices is similar to that of storing and selling any perishable item such as fish or fruit. It is very common for chip companies to have an "inventory reduction sale." So if you find yourself with inventory that has been sitting for some time, consider offering your customers products at special discount rates to move the materials quickly. Offering up to 50 percent discounts is common. (Inventory charges and scrapping have tax implications that are beyond the scope of this text. Working with an experienced financial expert or accountant to manage inventory is always a good idea.)

Order Management

Order management is another function that can be subcontracted to a semiconductor supply-chain management service. This may make sense, especially if you are resource-constrained or simply unwilling to manage the orders coming in for your product. In some cases, you may have only one product and will need to know some basics about order management, which starts when you receive a purchase order from your customer.

Acknowledge Purchase Orders

A PO that you receive from a customer requires a response from you to become a legal contract. Generally, a PO will be sent to you on-line or as a fax. You need to check the prices and volumes requested to ensure they are in line with your requirements, and then return a document to the customer with acknowledgment of the PO, which includes your committed date of delivery of the product. Note that there may be several delivery dates per purchase order for various quantities. Usually you try to line up your committed dates with the dates requested by customer. However, this is not always possible. Once you have sent the acknowledgment, the PO becomes a legal contract.

Order Supporting Material

When you start ordering material against acknowledged POs, double-check that supporting material orders such as lead frames, packaging substrates, packaging, and other supporting material is also ordered. Keep in mind that there are lead times for some of these items. During times of allocation these lead times can be significant. The last thing you need is a delay because you have to wait for lead frames or BGA substrate, which may be weeks in coming because you forgot to order ahead of time. It sounds basic, but it is surprising how even large IDMs forget to order these supporting materials.

Track Shipping

Also, keep in mind when you are ordering material that your subcontractors may not be linked to each other from a logistics point of view. For example, your foundry may not be linked to the assembly house. This means that the assembly house may not know for sure when your material is arriving from the foundry. You have to stay on top of your material as it is shipped from one stage of production to the next. As mentioned above, you should track the waybill and shipping numbers to ensure your products are moving on time.

If you are using a subcontractor to track these affairs, it makes sense to use a semiconductor supply chain company that can track all these waybill and shipping numbers for you as well as coordinate material flow communications between your vendors.

Terms and Conditions of Sale

Terms and conditions of sale (Ts and Cs) are normally attached to the invoice, which is sent out at the time of product shipment. (They are also attached to your acknowledgment of a purchase order.) The terms and conditions outline how you are to be paid for delivery of your product, as well as warranty provisions, liability limitations, and returns guidelines. The terms and conditions are fairly generic in the industry. However, it is best to consult with an attorney on these issues before you find yourself in a dispute.

Below is a list of typical topics explained in the terms and conditions of sale in the semiconductor industry.

- Payment terms
- Delivery terms
- Cancellation terms
- Taxes
- Warranty
- End of life
- RMAs
- Limitation of liability
- Jurisdiction (e.g., the State of New York, USA)

This list is probably the minimum that you will want covered in this document. Most likely you will want to specify more terms. For example, if your device requires a good amount of software that is either your property or third-party IP, you may want to add additional terms protecting the use of this software. If your device contains any technology that is restricted or requires licensing for export out of your country, you should ensure that your terms address this. If you do not intend your devices to work in certain applications, such as life-support systems or aircraft systems, you can specify this. Have your lawyer help in specifying your own terms and conditions.

Liability Limitation, Patent Indemnification, and Other Protections

Work with your lawyer to build liability limitations into your terms and conditions. Typically, a semiconductor company caps its liability to the dollar amount of chips sold. You will also need your lawyer to work out how your company needs to handle patent indemnification. And there may be other protections that would be appropriate in your particular situation. These topics require the help of legal professionals.

Payment Terms

You should be most interested in getting paid for your chips. The industry standard for payment is 30 days from the date of the invoice. Ensure that your terms specify the currency that you expect to be paid in. In addition, there may be instances where you want your customer to pay in advance. If your customer is a company you never heard of, in a foreign country, think about demanding cash in advance. You can also demand that payment be made by certified check, C.O.D., or through an irrevocable letter of credit (L/C). Once you have a track record with a customer who has always paid on time, you may decide to extend credit at a later date.

Cancellations

You also want to protect yourself from capricious cancellations. Your terms should spell out what is acceptable as a cancellation of your order. For example, you allow cancellation of a PO up to a certain time frame. You may specify that a PO can be cancelled up to 30 to 60 days before the delivery date, depending on how much custom work your products incorporate. If your product is a total custom chip for a customer, the terms of cancellation should be much stricter than those for a standard product.

Rescheduling

How you will handle rescheduling of orders is another issue that you may want to address. Once again, your terms should specify when and how purchase orders can be rescheduled. Rescheduling is normally allowed if it is requested within a certain time window before the delivery date.

RMAs

In semiconductor utopia, you never have to worry about Return Materials Authorizations (RMA). An RMA authorizes your customer to return parts to you due to failures in your product's form or function. Note the use of the word "authorization" in RMA. As your terms should specify, the customer will need your permission to return materials to you. This works in your favor (obviously) as well as the customer's, as it helps to track and account for these returns.

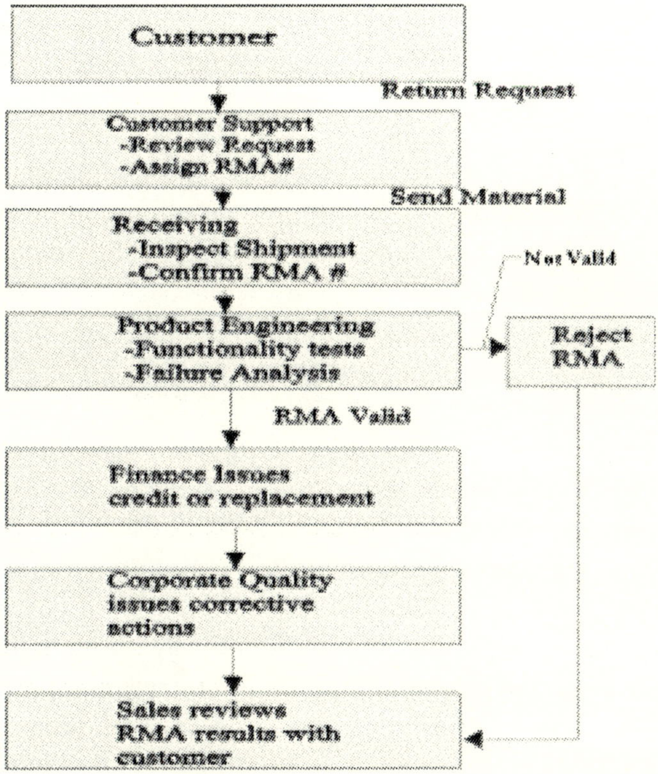

Figure 24: RMA Procedure

In the real world, RMAs are a fact of life. Murphy is always out there, and no matter how hard we try, the complexity and difficulty of delivering thousands of highly integrated complex systems that can fit on a pinhead will not always go smoothly. Especially with new products, many things can go wrong and customers may get defective products that they want to return to you for compensation or replacement.

In your terms and conditions of sale, you will address the extent of your warranty for your semiconductor products. Typically, product warranties extend from one to five years.

Depending on your specific market and business situation, it will be up to you whether choose to replace products outside of the warranty time-frame.

You absolutely need to institute a process for managing and tracking RMAs, in order to protect yourself and offer the best service to your customers. Also, because of the financial impact, proper tracking of RMAs is necessary to ensure proper accounting. As the numbers of your products and customers grow, you will need a process for tracing RMAs through your system.

The RMA flowchart in figure 25 outlines a typical process. For the start-up, no major investment in software is needed to track RMAs—just a good spreadsheet.

The first step in the RMA process is usually either an irate call from a customer directly to you or a frantic call from your salesperson in the field telling you there is a problem.

At this stage, it is important to get as much information as possible from the customer. Keep in mind that in many instances, failures in the field are not necessarily your fault, but may be due to mishandling by your customer. It is not uncommon that OEMs using your devices will have instituted incorrect production methods in their line that cause your chips to fail. Even famous OEMs with extensive production experience can make elementary mistakes. Faulty soldering procedures are particularly problematic. OEM system designs are also possible causes of chip-level failure. Even physical handling of devices can lead to problems due to static or simple blunt trauma to your chips.

Unless the cause of failure is obvious, you must work with your customer to find the root cause. Obtain samples of failing parts for failure analysis. It may be possible to ask your customer to send a failing system so you can inspect the problem. Insist that all returned shipments are properly packaged to avoid further damage to either the system or the chips. (Often customers forget to do this because they feel there is no reason to protect devices that are already broken.)

Once you have the material in your lab, you can begin failure analysis procedures. These procedures should move from the system level down to the chip.

The first step is to validate the problem. Try to recreate the problem either on your in-house test board or on the system provided by the customer. You may want to visit the customer site as well in order to understand the problem better. Some obvious questions to ask are

- Does the problem exist with every chip or a subset?
- Is the problem intermittent?
- Are any symptoms associated with the fault chip overheating?

Failure Analysis

Once you have validated a fault, you can begin your failure analysis (FA). From the high level, there are usually two phases to FA. The first phase tries to ascertain the root cause through functional tests. In this phase of the analysis, both system- and chip-level causes are considered.

If no external factors or obvious chip factors are found in the first phase of FA, the analysis proceeds to the next phase. This second phase concentrates on searching for the root cause in the internals of the chip. This is where special techniques such as current-voltage characterization, x-ray analysis, and scanning acoustic microscopy are used. If nondestructive tests do not reveal the cause of failure, destructive methods are used. Destructive tests include SEM analysis and laser or ion milling.

Once a failure is identified, corrective actions, if any, will need to be implemented to correct the problem in the process.

From the customer's perspective, any failure that is not due to customer handling or system implementation is subject to replacement or compensation. (This is why it is so important to include a limitation of liability in your terms and conditions.) If the root cause is found to be due to faulty manufacturing, you may seek compensation from one of your subcontractors. However, if the problem is a result of your design, you will have to pick up the expense of replacing faulty devices.

Major Points of Chapter 8

- The trick to production planning is to understand three basic concepts: your own company's demand forecast for your chip, the current yield, and the cycle time through the foundry.
- In setting up your planning in a database file, you should track at a minimum the customer request date for product, the projected actual date of delivery based on your calculated cycle time, and the foundry commit dates.
- When you are ordering material, keep in mind that your subcontractors may not be linked to each other from a logistics point of view.
- Use of a semiconductor logistics specialist can offer a great deal of help in tracking and managing the movement of your material through the manufacturing and shipping process.
- You will want to negotiate with a subcontractor to provide inventory for your finished products.
- A purchase order you receive must be acknowledged by you to make the PO a legal contract.
- Terms and conditions outline how you are to be paid for delivery of your product as well as warranty provisions, liability limitations, and returns guidelines.
- Institute a process in order to manage and track RMAs.

- In the RMA process, establish a phased failure analysis strategy.

CHAPTER 9

Cost and Pricing Analysis

One of the keys to any successful business strategy is to have a good understanding of the costs of doing business. The information in this section does not replace the need for a good accountant or financial manager to do the necessary detailed work to make certain your business remains solvent. The purpose of this chapter is only to highlight the salient costs associated with manufacturing a semiconductor product.

Several key manufacturing costs are typically considered "big ticket" items for an IC manufacturing project. Among these are up-front costs such as IP, mask tooling, prototype wafers, probe and DUT board development, and package and test development. Once in production, there are costs associated with wafer production, assembly, sorting, and testing. Also, in production you must pay per-unit package costs, which could get very expensive if you are considering a mammoth flip-chip.

This chapter will show some examples to illustrate how these costs can add up. The dollar values quoted below are typical in the industry at the time of this writing. Keep in mind that there may be other costs associated with your specific device.

Mainstream Cost Scenario

The first example illustrates the cost of a mainstream product. The device being manufactured is an integrated circuit targeted for a communications application. The device integrates analog-to-digital converters, memory, basic I/O, and an embedded processor. The device will be produced in the 0.18um process. The package is a 400-pin generic BGA.

The company is a start-up that does all back-end design in-house. A perfect candidate for the COT model, this company is working directly with a pure-play foundry and a one-stop assembly and test house.

Up-Front Costs

In order to get production rolling, the company needs to make several up-front investments, including IP licensing fees, development NREs, and mask tooling.

From the foundry, basic libraries are available free of charge. These basic blocks include some I/Os, memory compilers, PLLs, and ROM compilers.

The device also requires some higher performance blocks provided by the foundry's intellectual property partners. These blocks include a PCI I/O running at 33 MHz, a 10-bit ADC running at 20 Mbit/sec with nominal linearity MSMR, and a low-jitter PLL circuit. To obtain these blocks, the company pays straight license fees for one-time use, as shown in the table below.

In addition, the device requires an embedded processor capable of 250 MIPS throughput. For this block, the company negotiates a license fee of $400K plus a 4 percent per-unit royalty on the total end price of the chip. In addition, the company must pay the processor vendor an additional $100K in porting fees to have the process work in the targeted design and production environment.

Additional up-front fees are incurred in setting up manufacturing. The first major foundry charge usually encountered is mask tooling. As process technologies shrink, and the number of masks per device increases, mask tooling charges continue to skyrocket. Our example company, which uses the 0.18um process for their communications chip with five-layer metal, will typically pay around $180K for mask tooling charges. (Mask tooling can easily approach or exceed $500K for leading-edge technologies.)

For the prototype run, the foundry will generally run 12 wafers and guarantee that six wafers out of foundry will meet PCM specifications. Wafer costs vary with market demand and process technology. For example, in 0.35um the wafer cost can be as low as $1.3K, while in 0.15um you can expect to pay as much as $3.5K. Since our company is going for .18u, five-layer metal, dual-voltage, single-poly process, they are paying $3K per wafer or $36K for the proto-wafer lot.

Setting up the sort, assembly, and test incurs some nominal charges for services such as probe card and DUT board development. Costs incurred for these expenses are $5K and $2K respectively. Up-front test program development costs for the device amount to $10K.

The company in this example is targeting a generic 400-pin BGA package. Once again, nominal charges are incurred for PCB substrate development, which sets our company back $5K. These costs are summarized below.

10-bit ADC at 20 Mbits/sec	$ 50K
PCI I/O at 33 MHZ	$ 35K
Low-Jitter PLL	$ 35K
250-MIPS Embedded Processor	$ 400K
Porting Fees	$ 100K
Mask Tooling	$ 180K
Proto-Wafer Lot	$ 36K
Probe Card Development	$ 5K
DUT Board Development	$ 2K
Test Program Development	$ 10K
PCB Substrate Development	$ 5K
TOTAL UP-FRONT COSTS	$ 858K

Total up-front costs amount to $858K. This figure does not include smaller expenses such as shipping and inventory set-up costs. And the 4 percent royalty fee for the microprocessor will show up when calculating the final cost per chip.

Chip Production Costs

The next step is to see what our company is paying for production costs. In production, there are four per-unit costs: wafer production, assembly, sort, and test.

In this example, the cost per wafer for our company in 0.18um is $3K per wafer. If the company decides to start production with an order for 100 wafers, and there are 400 gross die per wafer (GDPW), the company will pay $300K for their production start in wafers.

Typical manufacturing will lose some wafers due to breakage or misprocessing. The resulting lots minus this loss are known as "in-line yield." For the purposes of this example, assume that there is no wafer loss in production, hence the in-line yield is 100 percent. There are 40K units total of testable dies.

Sort cost is calculated as a function of the time to test each die. For our example, typical sort times per die unit are around 2.5 seconds (not including set-up time), and sort times per wafer are roughly 12.5 minutes. As we discussed earlier, test costs are a function of the ATE test time per a particular device. In our example, assume that the tester costs $100 an hour. If it takes 1.25K minutes get 100 wafers through (12.5 minutes per wafer x 100 wafers), and sort testing is charged at $100 for every 60 minutes, the total price is $2083 for the wafer lot. Assuming that the defect density of the process is 0.2 defect per square inch, we can calculate the yield of the process at 80 percent.[1] Since we started with 40K die units and 20 percent fail, we are left with 32K die, which are sent to assembly, packaging, and testing.

Per-unit package costs can range from cheap to exorbitant. Since our example company decided to use a BGA with a relatively high pin count, they can expect to pay about $3 per chip

1 As discussed above, either the Bose-Einstein or the Murphy model can be used to estimate process yields based on defect densities. Your foundry can specify which formula it uses.

for packaging. For 32K chips, this totals $96K. In this example, we are assuming that assembly costs are included in the price. (You start to see why you want your sort test to be as comprehensive as possible.)

Package and leadscan will yield about 98 percent, leaving us with 31.36K chips. After packaging, the chips need to go through final testing. Here we assume a tester time of three seconds per chip. Final tests are generally a repeat of the sort test; we are assuming $100 per hour for final testing.

At final testing, the chips are generally tested at a higher temperature, and depending on the final application may be burned in to ensure no infant mortality. In our example, we assume that tester time is 3 seconds per chip including indexing. The total time to test 31.36K finished chips therefore is 26.13 hours, which costs $2.613K. The final test yield is assumed to be 98 percent. Hence, the number of chips available at the end of final testing is 30.73K units.

As shown in this example, final yield is a multiple of line yield times probe yield times assembly yield times final test yield. In our example, the final yield is 76.8 percent (100 percent x 80 percent x 98 percent x 98 percent).

The total manufacturing cost per chip is the sum of all the manufacturing costs divided by the total chips yielded. The costs, which include wafers, probe testing, assembly and packaging, and final testing, total $400K for about 30K chips. The cost per chip, therefore, is approximately $13.30 ($400K/30K).

Usually, a company will also add overhead to the per-chip cost. Overhead normally includes management, sales, and marketing charges. You can work with an accountant to establish these charges.

Finally, the price you should charge for your chip is a function of your market. In this example there is a royalty charge of 4 percent of the final chip price. Assuming that your final price per chip is $60, you will have to pay $2.40 per device in royalty. This cost must be added to the other per-chip costs prior to calculating your final gross margin.

Flip-Chip Cost Scenario

In the previous example, the communication device had an up-front investment cost of $858K, and a cost of about $400K for passing a production run of 100 wafers through the foundry. In total, the first batch of chips cost our start-up about $1.25M.

Let's use the above example to highlight some of the differences in cost one can expect for a flip-chip design. In the above example, the company decided to package its chip in a 400-pin BGA. Let's now say that they need a 1,200-pin flip-chip BGA and see how the calculations change.

Up-Front Costs

In addition to the up-front costs incurred above, there are additional design costs for flip-chips. These include the design of the redistribution layer (RDL), which is a metal layer that migrates I/O from the periphery of a design to the die interior. NRE for this design are typically in the $10K range.

There are also NRE associated with flip-chip package development. Flip-chip is only now becoming commercially available; therefore, most packages require custom level design. The NRE for flip-chip package designs are in the $30K to $60K range.

Production Costs

As mentioned above, the flip-chip process includes an extra RDL layer as well as a bumping process prior to packaging. At the time of this writing, RDL and bumping charges are each $100 per wafer.

Package costs are also higher than the previous example. The costs for an average package, such as a 37.5mm ceramic FC-BGA with 1,296 balls, are $45 to $60 per chip. (Note that all these prices may change radically over time.)

Test program development can be expected to be higher as well, since there will be some complexities associated with the high pin count. In addition, if care was not taken in the beginning of the design cycle to create a cost-cutting test strategy, the device will need to be tested on a much newer and higher-priced ATE. This means that test times per unit could easily increase by a factor of 2.5.

In total, from a design standpoint, you can expect at least an additional $40K to $70K in additional NRE for your flip-chip product. This figure does not take into account additional test development costs or general design costs which could increase this number significantly.

Redistribution layer (RDL) and bumping services account for only a slight increase in per-unit costs when chips are mass-produced. For example, in the traditional example, per-wafer costs were $3K. With additional flip-chip overhead for RDL and bumping services, the cost would increase by only $200. This increase would represent a premium of less than 8 percent in per-chip costs at final test.

The real hit is in package and test costs. The flip-chip package costs $30 to $60 per chip, compared with about $3 for the example above.

Major Points of Chapter 9

- Up-front costs include I license fees, mask tooling, prototype wafers, probe and DUT board development, packaging, and test development.

- The per-unit manufacturing cost is a combination of final chip cost, royalties, and your own overhead.

- To calculate chip cost, you divide the wafer cost by the number of chips per wafer, then add additional costs of assembly, testing, and delivery.

- The extra process steps required for flip-chip products can add cost to the final product when compared with mainstream products.

CHAPTER 10

Flip-Chip Considerations

On the high end of chip packaging designs is the flip-chip. Flip-chips have only recently come into play on the commercial scene, even though the technology has been around for almost 30 years. You can expect to pay higher prices for flip-chip technology, as we saw in the last chapter. But for those who have extremely high I/O and speed requirements, flip-chip may be the only commercial option.

Two industries in particular benefit a great deal from flip-chip technology: the networking industry, which requires chips with over 1000 pins per chip to accommodate routing activity, and the wireless industry, which seeks high-performance, low-power, mixed-signal applications with extremely small form factors. Obviously, these benefits will appeal to many other markets as well.

The concept behind flip-chip technology is fairly straightforward. Instead of connecting the I/O pads of a chip to an exterior pad ring, I/O connections are made directly from the face of the die to a substrate by means of a small conductive bump. The direct contact with the flip-chip eliminates the wire bond,

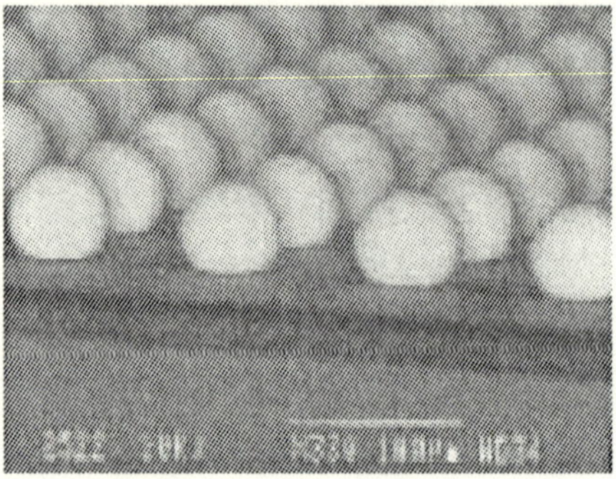

Figure 25: RDL Bumps (Courtesy of UMC Corporation)

which lowers inductance in the I/O path and improves electrical performance. Also, moving the I/O interconnects from the periphery to the center of the die means that you can fit more I/O interconnects per die than with traditional methods, allowing for smaller die sizes. It also means you can have much smaller packages than traditional methods allow.

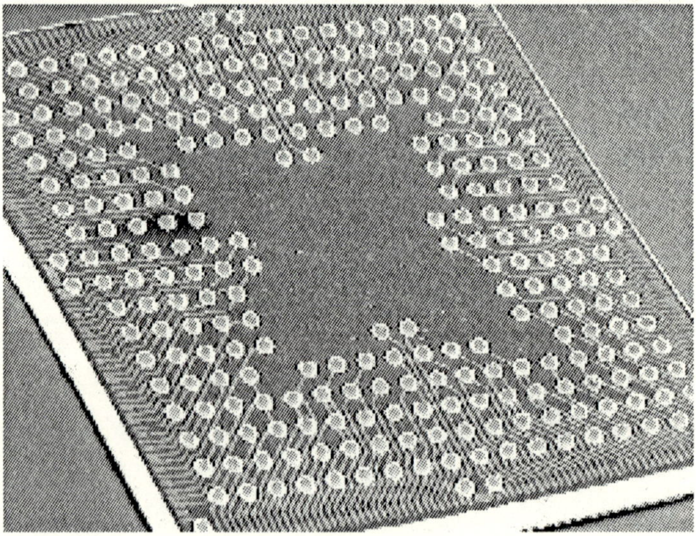

Figure 26: Flip-Chip Ball Out (Courtesy of UMC Corporation)

Redistribution Layer

Today, the I/O migrates onto the face of the die through the use of a redistribution layer (RDL) applied in the wafer manufacturing process. The RDL is a metal layer that reroutes I/O from peripheral pads into the die interior. RDL reroutes the I/O, clock, and power pads to the die surface in a dual-row, multiple-row, full-array, staggered-array, or other acceptable pattern. RDL allows ICs with peripheral pad pitches of <150um to be bumped and assembled cost-effectively. Currently, technology can place more than 6,000 I/Os on a die. In fact, the number of die-level I/O that can be placed using flip-chip is outstripping the capability of routing these I/O signals over a PCB substrate.

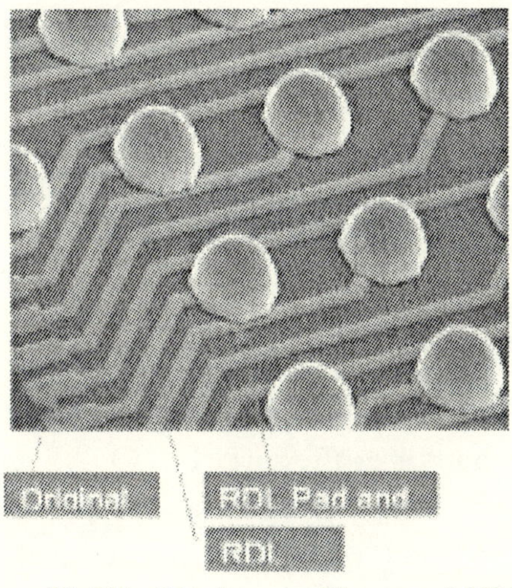

Figure 27: Flip-Chip Layout (Courtesy of UMC)

The industry is working to eliminate the need for peripheral I/O and the RDL, but since foundries today use peripheral

I/Os in their libraries, you will need to have an RDL for the time being. (The technology to provide interior core I/Os is still evolving. A few challenges include finding ways to provide ESD protection and metal routing across the I/O core.)

The flow diagram below shows how the RDL is generated and then bumped.

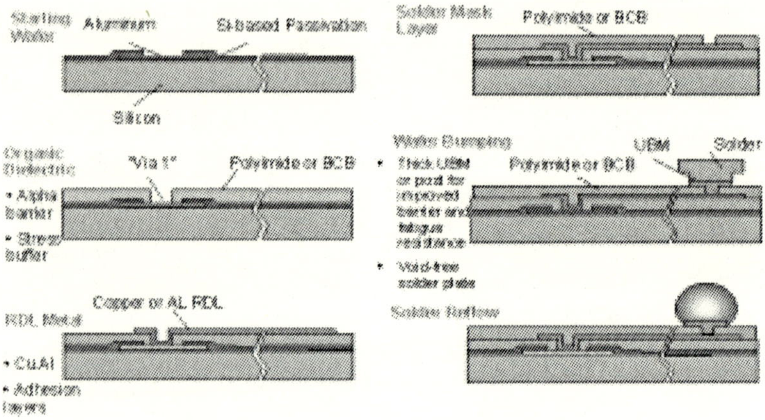

Figure 28: RDL Flow (Courtesy of UMC)

Bumping

Once the I/Os are available on the die surface, you need to apply conductive bumps to the surface I/Os which will interface with the flip-chip package. Paste-deposition and electroplating are two types of commercial bumping services that are available at major assembly subcontractors. These have been licensed from Unitive and Flip Chip Technologies, respectively.

The primary difference between these is in their approach to depositing solder on the IC wafer. Paste deposition processes apply a solder paste directly to the surface of a wafer that has been prepared with a photo-lithographically defined template Electroplating processes deposit solder using electrochemical deposition into a similar template. Paste deposition processes were the first on the scene and were originally used in automotive under-the-hood electronics. The plating processes, on the other hand, are currently used in high-volume applications such as high-end logic.

From a COT planning perspective, flip-chips introduce some new issues including planning complexities, longer lead-times, new chip and test-level design considerations, and extra expenses.

RDL/Bumping Vendor

Flip-chip manufacturing today also brings at least one extra vendor into the manufacturing process, the RDL/Bumping vendor. These vendors work with the major pure-play foundries today to provide RDL design and bumping services. Their up-front involvement with your chip design is critical to your success.

Flip-chip packaging adds a great deal of complexity to process planning. This is because almost every stage in the flip-chip development process has to be qualified prior to commencement of production. You must optimize the routing of your package before committing your netlist to the bumping vendor, to be sure that the array can be routed through the package and meet your timing and jitter requirements. This is a bit counter-intuitive, but it is critical because once you have bumped your wafer, you are committed to a certain I/O array on the face of your die. And the last thing you want is for some jitter issue to surface that requires a redesign.

So you should work closely with both your RDL/bumping vendor and your package house, and commit the netlist only after you and the package vendor agree on the substrate layout. The RDL/bumping vendor can generate an RDL mask design from the design's netlist data; the foundry will include this mask in the wafer build. To complete the flip-chip package design, the package vendor will need a pad-ball-mapping listing, a signal/power pin listing, and the netlist.

These extra steps mean longer cycle times for flip-chips than for traditional chips. For example, you can expect to need two to three weeks to optimize the substrate design; one to two weeks to review, tweak, and approve the design; and ten to twelve weeks for substrate development. That means lead times of 13 to 17 weeks just for prototype development.

A detailed diagram of the flip-chip process is provided below.

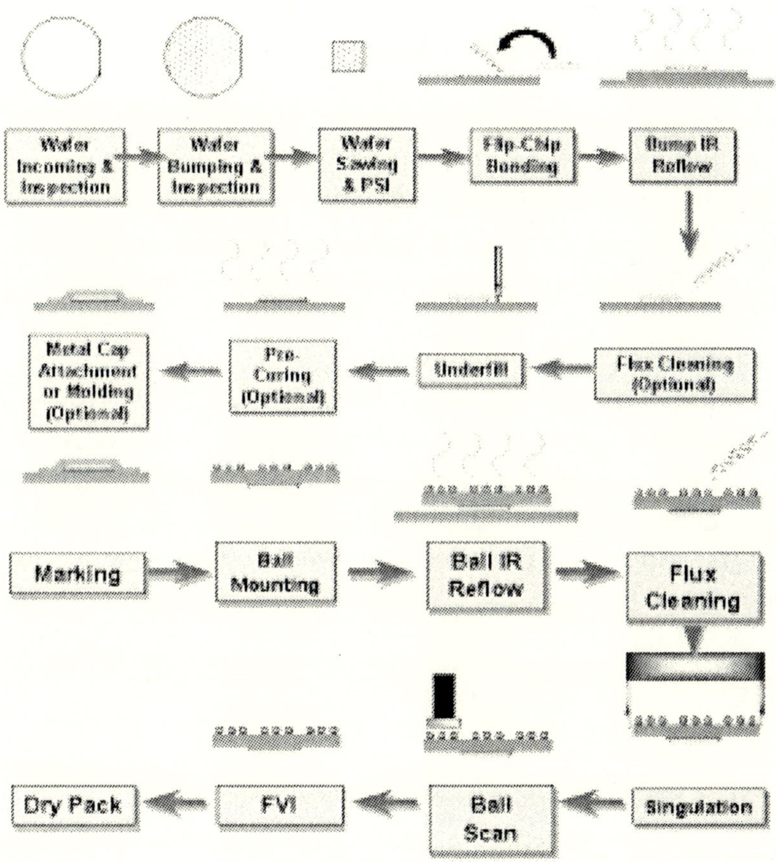

Figure 29: Flip-Chip Process Flows (Courtesy of UMC)

Failure Analysis and Testing

Flip-chips also present some new challenges for failure analysis and testing. Popular methods of failure analysis, particularly focused ion bean (FIB) techniques, become difficult with flip-chips because a flip-chip's RDL I/O layer covers the metal surface of the die, making FIB techniques quite difficult. (Traditional manufacturing puts I/O on the periphery allowing direct FIB access to the metal layers.)

Therefore, when designing a flip-chip, is it important to design with FIB failure analysis in mind. With a flip-chip it may be necessary to use backside FIB techniques, which require extreme accuracy and excellent mask alignment. It is a good idea to talk with a FIB vendor prior to committing to your design.

For testing, flip-chips can be probed before or after bumping. The data is not conclusive on which method is better. You should work with your test vendors to agree on a process. In any case, testing of flip-chips is expensive because of the high pin counts and speeds associated with flip-chip designs. Many of the strategies discussed earlier for reducing test costs can also apply to flip-chips; these cost-saving strategies take on more importance when your pin counts start exceeding 1000 pins.

In the last chapter, we saw how additional NRE and package costs add up in a high-pin-count flip-chip design. Flip-chip products are expensive because they involve extra design and manufacturing steps. Also, standard commercial methodologies for flip-chip manufacturing are still evolving. At the time of this writing, there are over 50 different ways to produce flip-chips. The good news is that more companies are coming online to support flip-chip technology. As these companies gain experience and increase in number, the costs charged for flip-chip designs will come down.

As flip-chip technology becomes commercialized, the industry will adapt to provide standardized manufacturing approaches as well as tools, testing, and failure analysis techniques to better support the technology. There is no doubt that flip-chip represents the next important technology trend in semiconductor manufacturing. As prices come down, more and more markets will benefit from the flip-chip advantages of pin count, speed, and smaller die and package sizes.

Major Points of Chapter 10

- Flip-chips have only recently come into play on the commercial scene, though the technology has been around for almost 30 years.

- Many new designs should consider flip-chip technology, especially designs that require extremely high pin counts or high speed in an extremely low package.

- Flip-chip packaging adds a great deal of complexity to process planning.

- Two types of commercial bumping services, licensed by Unitive and Flip Chip Technologies, are available at the major assembly subcontractors.

- Work closely with your RDL/Bumping vendor and your package house, and commit the netlist only after you and the package vendor agree on the substrate layout.

- When doing an array-bumped flip-chip, it is important to design for FIB.

- At the time of this writing, the typical cycle times for flip-chips are long.

Appendices

GoFabless.com

As we mentioned in Chapter 1, the goal of this book is to help you make your COT project a success. For additional help and information, please log on to www.GoFabless.com, a site that is dedicated to COT manufacturing. The site provides industry links, forums, and other sources of information for the COT planner. Also, if you have any questions, comments, or inputs, we would be more than happy to hear from you at www.GoFabless.com.

Fabless Semiconductor Association

The Fabless Semiconductor Association (FSA) was established to promote the fabless business model, as well as to create an open forum for building relationships between fabless companies and their suppliers. Established in 1994, the FSA has grown to include more than 350 members, including fabless companies, IDMs, and suppliers representing a broad range of semiconductor applications throughout the supply chain. The Association is a vital source of information for prominent media and analyst outlets across the globe.

The FSA's vision is that half of all integrated circuit revenue will come from fabless operations by 2010.

The FSA prides itself on the tools it provides to its members to accelerate the growth of the fabless model. Some of these tools are listed below.

- Annual surveys focused on foundry services, information technology, fabless salaries and compensation, fabless and IDM outsourcing, supplier offerings, and wafer pricing. The cornerstone study, the Annual Wafer and Packaging Demand Survey, is a vital tool used by foundries to achieve an optimal balance between wafer supply and demand.

- Directories and databases provide extensive profiles of industry players and various trend data on all aspects of the industry.

- Quarterly networking events openly discuss industry trends, provide access to financial and industry forecasting information, and promote education on key industry topics.

- Quarterly publications trumpet the successes of FSA members and provide thought-provoking articles on industry issues.

- An Annual Suppliers Expo allows Fabless companies to see suppliers' latest products and services.

- An Annual Awards Dinner celebrates the year's accomplishments and gives awards for outstanding achievements in the industry.

Additional information on the FSA and its members, publications, reports, and upcoming events can be found at www.fsa.org or by contacting Vivian Pangburn, (972) 866-7579, vpangburn@fsa.org.

Manufacturing Services and Vendors

Assembly and Packaging

Amkor Technology, Inc.: www.amkor.com
ASAT Inc.: www.asat.com
ASE Ltd.: www.asetwn.com
ChipPAC: www.chippac.com
Flip Chip Technologies: www.flipchip.com
IBM: www.ibm.com
OSE: www.ose.com.tw
Pantronix: www.pantronix.com
SPIL Siliconware: www.spil.com.tw
STATS: www.statsus.com

Burn-In, Test, and Failure Analysis

Accurel Systems: www.accurel.com
Amkor Technology, Inc.: www.amkor.com
Analytical Solutions: www.analyticalsol.com
Artest Corporation: www.artest.com
ASAT Inc.: www.asat.com
ASE: www.asetwn.com
Bridgepoint Technical Manufacturing: www.bridgept.com
ChipPAC: www.chippac.com
Credence Systems: www.credence.com:
Digital Testing Services (part of ISE Labs): www.dts.com
FIB International: www.fibinternational.com
ISE labs: www.iselabs.com
Pantronix: www.pantronix.com

Solid State Testing: www.sstmicro.com
SABER: www.slb.com/saber
SPIL Siliconware: www.spil.com.tw
STATS: www.statsus.com
Teradyne: www.teradyne.com

IP Providers

ARC Cores Ltd.: www.arccores.com
Artisan Components: www.artisan.com
ARM: www.arm.com
Chipidea Microelectronica, S.A.: www.chipidea.com
Dolphin Technology: www.dolphin-ic.com
inSilicon Corporation: www.insilicon.com
LEDA Systems, Inc.: www.ledasystems.com
Mentor Graphics: www.mentorg.com
MIPS Technologies: www.mips.com
MystiCom Ltd.: www.mysticom.com
NurLogic Design Inc.: www.nurlogic.com
Parthus: www.parthus.com
QuArc: www.quarc.com
Siroyan: www.siroyan.com
Sonics Inc.: www.sonicsinc.com
Synopsys: www.synopsis.com
Tensilica: www.tensilica.com
Virtual IP Group: www.virtualipgroup.com
Virtual Silicon Technology, Inc: www.virtual-silicon.com

Design and Prototype Services

IMEC-Europractice: www.europractice.imec.be

Tality: www.tality.com
Simplex Solutions: www.simplex.com
Time to Market: www.time2mkt.com
QThink: www.qthink.com

Shipping and Logistics Services

JSI Shipping: www.jsishipping.com

Photomask Services

DuPont Photomasks: www.dupont.com
Photronics: www.photronics.com
Taiwan Mask Corporation: www.tmc.com.tw

Pure-Play Wafer Foundries

1st Silicon: www.1stsilicon.com
Amkor Technology, Inc.: www.amkor.com
Chartered Semiconductor: www.charteredsemi.com
Dongbu: www.dsemi.com
Grace Semiconductor: www.gsmethw.com
Silterra: www.siltera.com
TSMC: www.tsmc.com
UMC: www.umc.com

Glossary of Abbreviations

A

ADC	analog-to-digital converter
AFM	atomic force microscopy
API	application program interface
ASIC	application-specific integrated circuit
ATE	automatic test equipment
ATPG	automatic test program generation

B

BE	Bose-Einstein
BGA	ball-grid array
BIST	built-in self test
BLR	board-level reliability
BU	build-up

C

C.O.D.	cash on delivery
CAD	computer-aided design
CAM	content addressable memory
CD	critical dimension
CIVVA	charge-induced voltage variation analysis
CMOS	complementary metal oxide silicon
COT	customer-owned tooling
CSP	chip-scale packaging

D

DAC	digital-to-analog converter
DBFIB	dual-beam focused ion beam
DFT	design for test
DIP	dual in-line package
DMA	direct memory access
DRC	design rule check
DSP	digital signal processing
DUT	device under test

DVD	digital video disc

E

EDA	electronic design automation
EDRAM	embedded dynamic random access memory
EMS	electronic manufacturing services
ERC	electronic rule check
ESD	electrostatic discharge
EVI	external visual inspection

F

FA	failure analysis
FC-BGA	flip-chip ball-grid array
FE-AES	field emission auger electron spectroscopy
FEC	forward error correction
FIB	focused ion beam
FPGA	field programmable gate-array
FSA	Fabless Semiconductor Association
FTP	file transfer protocol
FYI	for your information

G

GDPW	gross die per wafer
GPIO	general purpose input/output

H

HDL	hardware description language
HSTL	high-speed transistor logic

I

I/O	input/output
IC	integrated circuit
IDDQ	quiescent current
IDM	integrated device manufacturer
IP	intellectual property

J

JEDEC	Joint Electron Device Engineering Council

L

L and C	inductor and capacitor
L/C	letter of credit
L-effective	effective channel length
LQFP	low-profile quad flat package
LRC	inductance, resistance, and capacitance

M

MIF	mechanical interface
MPW	multi-project wafers
MQFP	metric quad flat package
mux	multiplex
Mvia	metal via

N

NDA	non-disclosure agreement
NRE	non-recurring engineering

O

OEM	original equipment manufacturer
OPC	optical proximity correction
OS	operating system

P

PBGA	plastic ball-grid array
PC	personal computer
PCM	process-control module
PDIP	plastic dual-in-line package
PLCC	plastic leaded chip carrier
PLL	phase lock loop
PMC	post-mold cure
PO	purchase order

PSM	phase-shift mask

Q

QAM	quadrature amplitude modulation
QFP	quadrature flat package

R

RC	resistance capacitance
RDL	redistribution layer
RF	radio frequency
RMA	return materials authorization
ROM	read-only memory
RTL	register transistor logic

S

SAM	scanning acoustic microscopy
SEM	scanning electron microscope
SIMS	secondary ion mass spectrometry
SMIF	standard mechanical interface
SO	small outline
SOC	system on a chip
SOI	silicon on insulator
SOIC	small-outline integrated circuit
SQUID	scanning quantum interference device
SRAM	static random-access memory

T

TEM	transmission electron microscope
TQFP	thin quad flat package
24/7/365	24 hours a day, 7 days a week, 365 days a year

V

VDD	supply voltage
VHDL	verilog hardware description language
VIL/VIH	voltage input low/voltage input high
VOL/VOH	voltage output low/voltage output high

W

WAT	wafer acceptance test
WIP	wafers in process

X

XPS-ESCA	x-ray photoelectron spectroscopy
XRD	x-ray diffraction

List of Tables, Forms, and Figures

Tables

Forms

Figures

Index

A

B

C

D

E

F

G

H

I

L

M

N

O

P

Q

R

S

T

U

V

W

Y

About the Authors

Anthony Simon

Anthony is the director for the digital cable set-top box IC product line for Conexant Systems, Inc., in San Diego, California. His responsibilities include P&L, product strategy, marketing, and IC program management for the product line. Anthony started his semiconductor career with VLSI Technology in San Jose, California (acquired by Philips Semiconductors). At VLSI, Anthony worked as a product marketing manager in VLSI's consumer digital entertainment division which developed standard as well as ASIC products for the set-top box market. With VLSI, Anthony also spent two years with VLSI's subsidiary Comatlas in Rennes, France. Anthony holds a B.S. degree in Electrical Engineering from the University of Akron and obtained professional engineering (PE) licenses in Florida and Virginia. Anthony also holds a Ph.D. in political science from the University of Debrecen in Hungary and a Master's degree in Economics from George Mason University in Fairfax, Virginia. Before his career in semiconductors, Anthony served in the U.S. Department of State as a diplomat; his service included a four-year posting to the U.S. Embassy in Japan.

Ronald Vogel

Ron has spent 15 years in the foundry and ASIC industry. During his career, Ron has worked in process engineering, test product engineering, planning and logistics, marketing, and sales. Ron joined VLSI Technology (acquired by Philips Semiconductor) as a Process Engineer and held various management positions in Product/Test Engineering and Product Marketing for communications consumer product divisions.

Ron was also responsible for creating and developing the Foundry Division at VLSI Technology and is currently responsible for new business development with UMC. Ron graduated from Arizona State University with a B.S. in Electrical Engineering.

Printed in the United States
5787